Czechoslovak Academy of Sciences

Aphid parasites

of

Czechoslovakia

Czechoslovak Academy of Sciences

Scientific Editor Ing. Ľudovít Weismann, CSc.

Scientific Adviser RNDr. Josef Šedivý

Language Editor Bessie Kloučková

Aphid parasites of Czechoslovakia

A review of the Czechoslovak Aphidiidae (Hymenoptera)

Petr Starý

Springer-Science+Business Media, B.V.

1966

ISBN 978-94-017-5225-1   ISBN 978-94-017-5223-7 (eBook)
DOI 10.1007/978-94-017-5223-7

Originally published by Dr. W. Junk, Publishers — The Hague in 1966

Softcover reprint of the hardcover 1st edition 1966

This book is dedicated to my teachers

| Prof. Dr. JULIUS KOMÁREK | corr. member of the Czechoslovak
Academy of Sciences

and

| Prof. Dr. JAN OBENBERGER | DSc., corr. member of the Czecho-
slovak Academy of Sciences,

The Charles University, Prague

Contents

8

I.

Introduction

Propagation of natural enemies as one of the means of controlling insect pests is a persistent problem of modern entomology. Many cases are known where insect pests were successfully controlled by natural enemies. Nowadays we know many effective natural enemies, and the research is continually being intensified and extended to other groups, the employment of which might be advantageous.

Recent progress in the taxonomy of parasitic *Hymenoptera* has been rapid. The modern conception of taxonomy, built on the conception of the biological species, is the basis for work in the ecology of parasitic insects, insect control, etc.

Complete co-operation, i.e. team work, on the problems of insect control, including the work of a taxonomist, laboratory and field ecologist, insect control specialist, etc., is an ideal manner of work. In addition, the study of parasitic insects necessitates close co-operation of people working in the branches mentioned in host-and-parasite groups, and only such cooperation enables the successful employment of a certain entomophagous insect in insect control.

Many recent examples prove that the isolated work of individual scientists, without contact with other specialists, may remain without any effect. For example the introduction of the "imported", but in fact native parasites, that were introduced in consequence of a wrong taxonomic determination of the entomophagous insect, is such an example. Sometimes the entomophags are introduced while the environment and host specificity range are disregarded — this cannot lead to success, but only discredit the biological control of insects.

In practice the complex team solving of problems is often difficult, as the specialists work in different institutes and in different countries, etc., in short, there are many objective obstacles in the work on a problem. It requires great international effort, and especially good will in the present as well as in the future. In practice these relations are solved by establishing big organizations of a more or less international character, that are continually being improved and extended (C.I.B.C., O.I.L.B., U.S. Dept. of Agriculture, etc.). The future demands a united international organization if the biological control of insects is to be successful.

The research of entomophagous parasites as well as their employment represent a very broad field of action. The taxonomists of large groups are so laden with work that they can hardly give more than basic information about certain species. On the other hand, the taxonomists of smaller groups are not of such a broad outlook, but their knowledge is more profound, including taxonomic research as well as the application of the entomophag in the insect control.

The author of this paper, originally a taxonomist and then a synecologist, apprehended the essential difficult problem of standpoint — whether to work from the standpoint of method or object. After valuable consultations with many specialists he endeavoured to give in this book the results of a few year's work on aphid parasites, from the taxonomic research, life-history of the parasites, host x parasite relationship, to their application in insect control. The author is well aware of the imperfection of the book in the chapters where the aspects of the problem, on which he has worked himself, predominate. That is the inevitable result of mono-authorship. He wanted to show at least a scheme of such work, of course written on the team — work basis, so that basic comprehensive works, broadly applicable, could be obtained.

The author is indebted to all specialists who have helped him in his work. They are: M. Achvlediani (Tbilisi), M. Ataeva (Dushanbe), †Dr. E. Bajári (Budapest), Prof. Dr. L. Brundin (Stockholm), Ing. I. Ceianu (Budapest), A. Davletshina (Tashkent), G. Dmitriev (Kiev), Prof. Dr. I. Docavo Alberti (Valencia), A. A. Dzhibladze (Tbilisi), R. D. Eady (London), Dr. H. H. Evenhuis (Wageningen), Dr. M. Fischer (Wien), Dr. D. Gerling (Albany), Prof. Dr. A. Goidanich (Torino), Dr. Ch. Granger (Paris), A. Gullyev (Askhabad), Prof. Dr. J. Györfi (Sopron), M. S. Hassan (Giza), Dr. E. Haeselbarth (Göttingen), Dr. K. J. Heqvist (Stockholm), Dr. O. Heikinheimo (Tikkurila), †W. D. Hincks (Manchester), O. M. Ivanova-Kazas (Leningrad), O. I. Ivanovskaya-Shubina (Novossibirsk), Prof. Dr. A. Kurir (Wien), Prof. Dr. C. H. Lindroth (Lund), †A. N. Luzhetzki (Tashkent), Dr. M. Mackauer (Belleville), Dr. M. Markkula (Tikkurila), Dr. P. Marsh (Davis), Dr. C. F. W. Muesebeck (Washington), Prof. M. N. Narzykulov (Dushanbe), Dr. D. Rosen (Rehovot), A. Rupais (Riga), A. W. Stelfox (Dublin), Prof. Dr. U. Sedlag (Dresden), C. F. Smith (Raleigh), Prof. R. D. Shenefelt (Madison), Dr. B. C. Subba Rao (New Delhi), Dr. M. Sorin (Osaka), Dr. E. I. Schlinger (Riverside), Dr. M. T. Tanaka (Utsunomiya), Dr. E. Tremblay (Portici), Prof. N. A. Telenga (Kiev), V. Tobias (Leningrad), Prof. Dr. P. Vukasović (Novi Sad), Dr. S. Wiackowski (Skierniewice), Prof. Dr. C. Watanabe (Sapporo), Prof. Dr. K. Yasumatsu (Fukuoka), Prof. Dr. A. A. Yakhontov (Tashkent), Dr. H. Zwölfer (Delémont).

Dr. J. Holman was kind enough to identify numerous aphid material and suggest valuable aphidological notes.

The author also sincerely thanks Mrs Štysová, Miss Sobotková and Miss Mikulecká for valuable technical assistance.

Prague, April 1964

II.

Methods
of collecting,
preservation
and identification
of material

Aphidiids are common parasites of aphids, and that is why they occur, like the aphids, in almost all kinds of terrestric habitats. Large numbers of aphidiids can be obtained by the sweeping method. This method has many disadvantages, first of all that the swept material is more or less deprived of environment, and it is necessary especially in parasitic insects to know the ecological relationship, i.e. the host, its host plant, the hyperparasites, etc., in short, the complete food chain; only thus, set research of the entomophagous parasites in the field is of some value at the present state of research. In addition, it is difficult to determine the swept material, because at random we get many specimens without knowing the variability range of the species, etc.; it is essential for further work on the newly described species from the swept material: any taxonomic confusion (biological races, etc.) cannot be solved, as with the exception of the habitat, we do not know its other relations to the environment. We meet with more and more such problems in modern taxonomy.

For previously mentioned reasons the best method of obtaining the material of the aphid parasites is the laboratory breeding from the aphid colonies taken in the field. The samples are numerous according to how much material we need. From a few days collection in the field, when on one day many and many tens of samples are taken, naturally the material requires much space, if taken in large bottles. From our experience, the sampling of the aphid colonies in small vials of 25 × 60 mm, used for medicaments, is most practical. The vials are thick-walled and resistant to mechanical damage, at their top there is a small groove (Fig. 1). The vials are covered with a piece of dense silon texture, tied with rubber; the groove is useful as the rubber cannot slip down. The parasites and aphids cannot escape through the silon texture, but have sufficient air. Only small parts of plants with the aphids should be put in the vials, not more than 2/3 of the vial — especially when the plants are succulent and soon mould. Each sample is numbered. The numbers of samples can be serial, with the number of year to prevent a mistake:

13

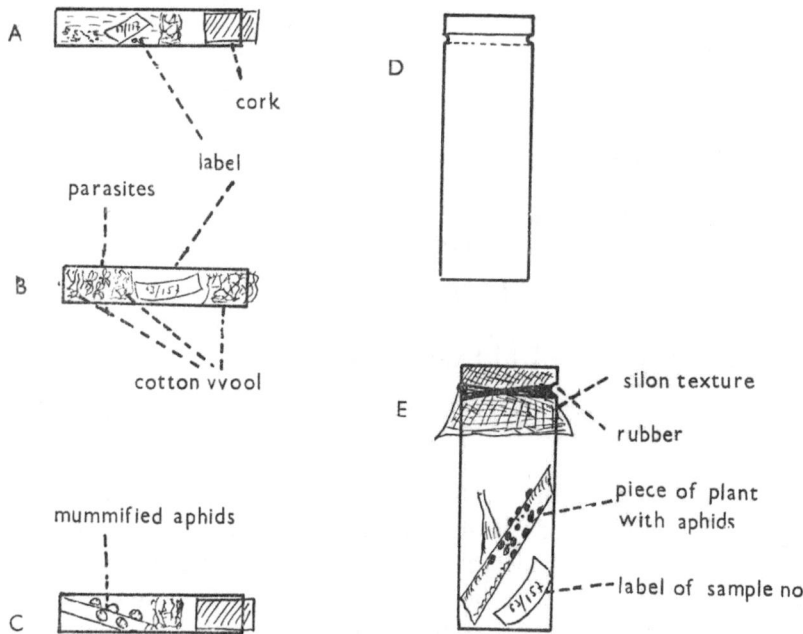

Fig. 1. Vials for collecting and preservation of aphid parasites. A — vial with aphids used for preserving field samples. Aphids are preserved in alcohol and pressed a little by a piece of cotton wool to take off bubble of air and prevent their quick passive movements and destruction. B — vial with parasites adults (dried), which are placed between two pieces of cotton wool to prevent their breakage by movements. C — vial with mummified aphids as used for collection purposes. D — vial used for samples of aphid colonies and breeding of parasites in the field. E — ditto, containing a piece of plant with aphids, and covered by silon texture.

e.g. 1963/159. A smaller number of aphids, mostly adults, is put in alcohol (this can be done additionally, in the evening). The plants are taken for a herbarium, so that later they can be determined in detail. Regarding the ecological relations, the aphid attending ants ought to be taken too. If the sample is larger, it is preferable to divide it into a few small vials to prevent it from moulding. The aphid colonies taken in the vials usually provide a sufficient number of parasites for observation. If we want to take a large sample, it is necessary to have a few reserve big plastic bottles with a silon cover, where long pieces of plants are put together with strips of filter paper; in the laboratory the material should be soon transferred to more spacious rearing cages or bottles.

If material for mass rearing is to be obtained, numerous aphid colonies are placed in big 5-litre-bottles, the bottom of which has been cut off. Crumpled paper or a net is put under the parts of plants, so that the material is aired from below and does not mould. Thick leaves, e.g. of sugar beet are hung in the bottle from above and freely aired.

The parasites in small vials are left to emerge and die, to avoid their immaturity. In larger breedings, where the moulding of material is possible, we prefer collecting the parasites with a suction-collector, as there is danger of their damage or loss in the mass of plant material.

The material in vials and laboratory breedings is revised a few days after the return from the field excursion, and the moulded plants are removed. Otherwise the material in the vials is left intact for at least a month from the day of sampling, as long as all primary and secondary parasites do not hatch. Then once more the material from plant remainders and mummified aphids is carefully revised, as often there are diapausing cocoons of parasites which hatch after several months, and only under certain conditions.

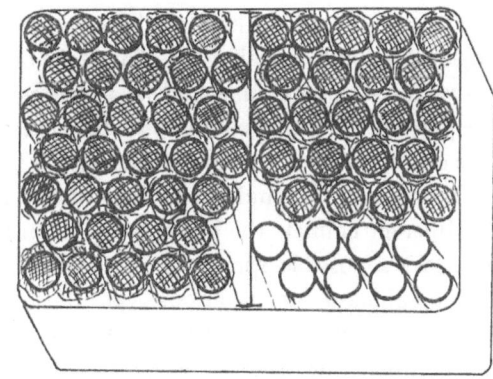

Fig. 2. Lower part of a field box containing vials with samples taken in the field.

A part of the mummified aphids is left for collections (Fig. 1) and transferred to smaller vials for possible revision of aphid identification, and for the coloration of the mummified aphids — this coloration or mode of pupation often enables at least generic identification of the primary parasite in the case that only the secondary parasites have been reared (see below).

The analysis of the material reared is carried out as follows: the piece of plant is carefully taken out of the breeding vial and the dead hatched parasites are dropped on a sheet of paper; there the primary and secondary parasites are sorted as well as the other natural enemies of aphids (*Chamaemyidae, Itonididae,* etc.) — these occur in mass in the breeding of the aphidiids. Each vial is labelled with the same number as the sample. The aphid parasites are seized with a fine pincette, preferably by the wings, so that damage is avoided, and put between layers of cotton-wool in small vials for later examination. The cottonwool may not be pressed too much nor too little in order not to damage or move the parasites during transport (Fig.1).

Searching for aphids and parasites. The aphids for the breeding and study of the parasites are not collected accidentally, but planwise with regard to the ecological data of the parasites we are desirous of obtaining. These lines should be observed:

1. The colonies of all aphid species in the habitat are taken (unless there is a homologous growth of a monoculture strongly infested by aphids).

2. Different types of habitats are chosen to obtain a basic knowledge about the fauna as soon as possible.

15

3. In each type of habitat the environment is classified, and the aphids from adjacent habitats are taken too. This is the only way to obtain a picture of the relations between the aphidofauna and its parasites in different habitats, food specificity of the parasites, etc. the basic knowledge necessary for the biological control of aphids. The classification of the aphidofauna from the adjacent habitats is necessary specially in the study of the aphidofauna of cultures. In the fields of monocultures neighbouring on different habitats the samples are taken in the border and middle of the field to ascertain a possible spread of species in the intermediate habitats.

The aphids occur in various types of habitats, as mentioned in the introduction. Mountains (alpine meadows, woods) are comparatively poorest in the aphidofauna and parasites: the lower the elevation is, the more the aphids occur. They are abundant on the borders of deciduous and mixed woods, in gardens, parks, orchards, and most abundant in meadows, balks and cultivated fields. The largest number of aphids, both quantitative and qualitative, can be found in Czechoslovakia in May and June in steppe districts (southern Moravia, southern and eastern Slovakia). Certainly the seasonal succession of species must be considered.

Special methods of sampling with regard to the life history of aphids are omitted, as the basic knowledge of methods of the searching for aphids is presumed. On collecting the aphids the damage caused by them on plants should be taken down, as in connection with the specificity of parasites the gall-producing aphids, root aphids and leaf-curling ones are attacked by various parasites; it should be considered that the ecology of one aphid species can vary with regard to the life-cycle. Ants are often a good indicator of the occurrence of aphids on plants, and they should be watched, too, with regard to the occurrence of parasites. Sometimes it is useful to tap the aphids down on a small plastic saucer (some mobile and quickly reacting species).

Besides the sweeping and laboratory rearing of aphids the parasites from peat-bogs can be obtained by using the Tullgren-apparatus, as the aphids living in peat-bogs are difficult to collect. Similarly the myrmecophilous aphidiids are obtained by the sifting of the ant-nest winter material, similar to other myrmecophilous insects.

Mounting. The most suitable manner of mounting the parasites is to stick the specimens on white standard labels 7 × 15 mm by the right side of the thorax. The wings of the stuck specimen are placed upwards, the abdomen is bent down, so that the characters on the propodeum and the lst abdominal tergite are well visible. This manner of mounting has many advantages:

1. Margins of the label protect the specimen quite well against mechanical damage (it is preferable especially to the sticking on the tip of a triangular label).

2. All parts necessary for determination are visible.

3. The white ground of the label is a good reflecting surface, improving the contrast of the picture of the object studied.

16

The specimen is stuck on the upper fourth of the label; on the lower margin the sign of sex and the number of antennal segments are written. At least 10 specimens should always be mounted to ascertain the basic data of the range of variability. The vial with the plant remainders and aphid mummies is left with the series of specimens.

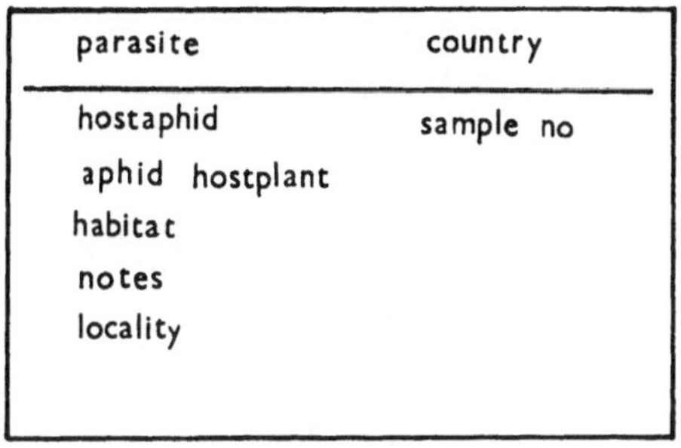

A file of breedings is kept in the laboratory in the series by years (63/150, 63/151, 63/152, etc.). The file card contains: locality, habitat, aphid host plant, aphid, notes (ants, etc.), i.e. records from the field diary (Fig. 3). The plants, aphids, etc. identified in addition are put down in the file later.

The first specimen of the series is provided with all data according to the file card, including the number of the breeding for better orientation in the collection.

For the examination of the characters on external female genitalia it is necessary to make slides, as follows:

1. In a dried specimen the apical part of the abdomen is carefully removed.

2. The specimen is given the same number as the slide.

3. The removed part of the abdomen is boiled in 10% KOH (as the thickness of the object requires).

4. The object is washed in water.

5. The slide is mounted in de Swann or in de Faure mounting fluid. It is useful to mount the right wing and perhaps other parts examined, e.g. antennae, etc. on one slide for later photographs.

6. The slide is given the same number and data as the specimen.

Fig. 3. Parasite and host catalogue card.

III.

Morphology
and
anatomy

Egg

The eggs of the *Aphidiidae* are of microscopical dimensions. The shape and size of an egg is specifical. The ovarian eggs are lemon-shaped or spindle-shaped, prolongated, etc. The dimensions of eggs in various species are e.g. 86 × 36 μ, 70−90 × × 24−29 μ, etc. The superficial membrane forms the chorion, with a small micropyle on the top. The dimensions of eggs that were laid in a host multiply their size several times. (Plate II, Fig. 9).

Larva

There are three larval (Plates I, II, Figs. 4, 8, 10−14) instars in the *Aphidiidae,* which may be easily distinguished by the shape of head, presence and shape of cauda, etc. The first instar larva has a distinct head and 13 body segments. It is remarkably mandibulate. The body segments bear various pubescence — either scattered hairs or rows of hairs, etc. The last segment bears a process that is called cauda, which is covered by hairs or spines. The cauda is mostly relatively short, to a lesser degree it is long and arcuate. Besides cauda, two lateral prongs are often developed, being perpendicular or oblique to the cauda. The second instar larva has the oral lobes developed in a different way. The mandibles are small. Pubescence on abdominal segments is lacking. The cauda is much shorter but always distinct, accessory prongs are lacking. The third instar larva is rather variable as to size. It may be characterized by the following characters: It is mandibulate, the cauda is lacking. The integument is slightly papilliform, the mouthparts are dorsal, there are spiracles developed on the body segments. The larva is rather mobile, especially when spinning a cocoon. Inside the last instar larva the oesophagus, mesenteron, salivary glands, malpighian tubes and rudiments of the reproductive organs may be recognized. The tracheal system is also recognizable. The spiracles are very small with a long closing apparatus.

Praepupa

In the praepupal stage the larva becomes shorter, the differentiation between segments and lateral folds is more apparent and further bends become recognizable on the body. Whitish fat bodies are visible through the skin (Plate XI, Figs. 5, 6).

Pupa

The pupa is of an exarate type, the segmentation is very fine. The legs are bent and pressed to the body. Antennae are laid along the ventral side of the body. Wings are folded and cover the lateral sides of metathorax, abdomen and the legs.

Imago

The body consists of the head, thorax and abdomen.

The head is orthognathous, bearing eyes, three ocelli, antennae and mouthparts. Eyes are of various shape and size, hemispherical, oval, more or less prominent, etc. The mouthparts consist of the labrum, mandibles, labio-maxillar complex. Mandibles are bidentate. Antennae are filiform or moniliform, with a various number of segments (10−30). The first antennal segment is called the scape and it differs distinctly by its shape and size from the following subglobular pedicel − the second antennal segment. Behind the pedicel, further antennal segments follow, called flagellum. Flagellar segments are usually of more or less the same shape and size; sometimes the middle segments of flagellum differ from the others. The apical segments differ by their shape from the preceding ones, they are often more or less distinctly united with the praeapical segment. On the head we may further recognize: the face, which forms about the frontal part of the head. Its part above the antennae is called the frons, its lower part is formed by the clypeus that covers the greatest part of the labrum and it is separated more or less distinctly from the face by a furrow. The clypeus is usually oval, more or less pubescent. On the sides of the clypeus there are tentorial pits that indicate the joining of a part of the endoskelet of the head. The top of the head is formed by the vertex which is separated from the occiput by the occipital carina. The occiput is declivous to the foramen occipitale. The part behind the eyes and under them is called gena. Its upper portion between the hind margin and occipital carina is called the temple (Plate I, Figs. 1−3, 5−7).

Thorax. The thorax consists of three main parts − the pro-, meso- and metathorax. These three parts are joined by the propodeum, that is phylogenetically the first abdominal segment. The prothorax is formed by the pronotum which is fused with the mesonotum at the upper side, the propleurae are comparatively weekly joined at its lower side. The pronotum is mostly smooth or slightly sculptured. The prothorax bears a pair of fore legs. The second part of the thorax is the mesothorax. Its dorsal part is called mesonotum, which is further separated into two main sclerites — the mesoscutum and the scutellum. The mesoscutum is often strongly developed and raised when viewed from the side or it is gibbous with an almost vertical declivity or gradual declivity to the prothorax. It is usually smooth and more or less pubescent, seldom almost granulate. The mesoscutum often bears two furrows — the notaulices. The notaulices may be distinct at the base only, or they may be distinct as far as the praescutellar groove or they may be missing. Their sculpture is various, usually they are more or less crenulate, rugose or almost smooth. The mesoscutum is separated from the following sclerite — the scutellum by the praescutellar groove that is of various depth and width. The scutellum is usually triangular, sometimes more or less prominent above the level of the mesoscutum, its margins are often crenulate, the surface being smooth, sparsely haired, rarely granulate. On the sides of the scutellum there are smooth lateral discs — the axillae, which belong to mesoscutum as to the morphology. The lateral parts of the mesothorax are distinctly divided downwards and they are called the mesopleurae. Their lateral sutures are usually crenulate, they are rarely slightly sculptured in the upper portion below the wing or they are weakly granulated. The mesothorax bears a pair of fore wings and a pair of middle legs. The metathorax: The dorsal part of the metathorax is formed by the metanotum. In its apical part there is a more or less visible tubercle — the postscutellum. On its sides there are flat, usually smooth or slightly sculptured impressions. The lateral part of the metathorax is not further divided and is called the metapleurae. They are usually smooth, slightly granulate to rugose, sparsely haired. They bear a pair of spiracles. The metathorax bears a pair of hind wings and a pair of hind legs. The propodeum bears a pair of spiracles. It has a conspicuous sculpture and pubescence. It is smooth or more or less rugose — often various carinae are developed, which divide the disc of propodeum into several smaller discs — the areolae. On the propodeum the abdomen is joined.

Wings: There are two pairs of wings developed in the *Aphidiidae*. There is one exception only — the female of *Diaeretellus ephippium* (Haliday) is wingless. The wing venation is either almost complete (in the rank of a family) or it is more or less reduced. The venation in the genus *Ephedrus* is the most complete. Fore wing: On the fore margin there is a strongly sclerotized fused costal and subcostal vein (and probably the radial vein, too) which dilates at the apex as the prostigma. The pterostigma reaches by its prolongation — the metacarp the wing apex. From pterostigma the radial vein extends, being composed of three abscissae and reaches the wing apex where it joins the metacarp and completes the pterostigmal cell in this way. Under the fused costal and subcostal vein there is the basal cell, being margined on the external side by the basal vein and on the lower side by the cubital vein which is developed towards the wing apex. Between the radial and cubital veins lies the median vein, which originates in the basal vein and points to the wing apex. It consists of three abscissae. On the upper side of this vein there are radial cells, separated from each other by two interradial veins. On the lower side there is the median cell, separated on the external side by the intermedial vein. Under the cubital vein there are two cubital cells, separated from each other by a transverse vein — the nervulus and by the anal vein on the lower side. Hind wing: On the upper margin there is a short fused costal and subcostal vein that deviates after a certain distance from the wing margin and prolongates as the subcostal vein. The radial vein is almost not developed and it is as distinct as a small point only on the lower side. Under the costal vein and subcostal vein there is the cubital vein which restricts the basal cell with the above mentioned veins, which is closed on the external side by the basal vein. Under the cubital vein there is a remainder of the anal vein visible at the fore part. There are many types of reduction of venation, which may be recognized from the key to the genera and species.

The wings are mostly hyaline, to a lesser degree they are smoky or bear dark spots.

The articulation of wings to the thorax. The wing articulates with the thorax by two processi of the notum of the given segment — with the fore and hind notal processi. In the lower part it articulates with the pleural process. The lower part of the wing membrane has the appearance of a ligament and it is called the axillary cord. Around the wing base there are situated the following sclerites — pteralia: Tegula, humeral plate, four axillary plates (acropteral, propteral, mesopteral, metapteral). The acropteral plate articulates with the base of the costal and subcostal veins, the propteral plate with the base of the cubital vein and by the intermediary sclerite also with the base of the anal vein. The lower part of the anal vein articulates with the mesopteral plate.

Modification of wings. Only a single case of wing modification is known in the *Aphidiidae* and it is a case of secondary modification, that is obtained during life. It occurs in *Paralipsis enervis* (Nees) and it was primarily supposed to be a primary adaptation of the parasite to the life in ant nests. Females of this parasite attack underground aphids attended by ants, which keep the aphid parasites as symphils and nibble a part of their wings.

Legs. The legs of the *Aphidiidae* are mostly slender, relatively long. A leg consists of the following parts: 1. Coxa, 2. trochanter, 3. femur, 4. tibia, which bears two spurs at its apex, 5. tarsus, the first segment of which is longer and called basitarsus, 6. praetarsus, bearing two simple claws, between which there is a small arolium. There is no special apparatus developed on the legs except the first pair which bears a cleaning apparatus of a tibio-tarsal type.

A b d o m e n. It is either rounded or lanceolate. The shape may be of sexual dimorphic character and then the abdomen is lanceolate in females and rounded in males.

21

The first abdominal segment is in reality the second one but for simplicity it is kept as the first one. In different segments the tergite and sternite, unified by a membrane, may be recognized. Both tergites and sclerites are comparatively weakly sclerotized. This feature enables great mobility of abdomen. 6 pairs of spiracles are developed in the abdomen. The first tergite has a varied sculpture, shape and pubescence. The spiracules are situated on its sides, being placed on more or less prominent spiracular tubercles. These spiracular tubercles are in some genera called primary tubercles, as, in addition, secondary tubercles that do not bear spiracles, are developed. All the abdominal segments are freely connected with each other except segments 3 and 4 which are fused, but the fusion is flexible. The genital segments of the abdomen bear, as in other *Hymenoptera,* the external genitalia.

External genitalia of the female. The ovipositor consists of appendices of the primary eighth and ninth abdominal segments. The base of the ovipositor form the first and second pair of valvulae, which are gonapophyses of bases of primary extremities of segments 8 and 9. The ramus of valvula 1 connects with valvifer 1 that is the base of urit 8. The ramus of valvula 2 is attached to valvifer 2, which is probably the coxit of the primary urit 9. The first valvifer, although it belongs to urit 8, is connected with abdominal tergum 9; the second valvifer articulates with the first one. With the distal end of each second valvifer the valvula 3 is connected (gonostylus). A pair of valvulae 3 is the sheaths of valvulae 1 and 2. In this paper valvulae 3 are called ovipositor sheaths. They have various shapes and pubescence, usually being covered with sparse hairs, to a lesser degree densely pubescent.

External male genitalia. They are formed by gonopods of the primary segment 9 and by a penis. The greatest sclerites of these organs are gonocoxits. On their latero-apical margin the gonobase is attached, which is probably formed by the fused basal parts of primary gonocoxopodits. On the internal ventral side of each gonocoxit a sclerit called volsella is attached. Its external prong is called cuspis, the internal one is digitus. To the distal end of each gonocoxit the gonostylus is attached. All the mentioned part is called gonoforceps. The penis is protected by the sheaths – penis valvae, which are primarily the basal processes of gonocoxopodits that separated themselves during development.

Modifications of abdomen. A special modification of abdomen may be found in the female of *Protaphidius wissmannii* (Ratz.), where tergites 1, 2 and 3 are normal, only tergite 3 beirg strongly narrowed at the apex. Beginning with tergite 4 the following tergites of the abdomen are tubiform, and form in this way a kind of sham-ovipositor. The external genitalia are situated as usual.

In some highly specialized aphidiid genera various accessory apparatus, enabling a better attack of the aphid, are developed. In general, two kinds of apparatus of this kind may be recognized in the aphidiids, being a clear example of convergence as to the development.

In the genus *Trioxys* 2 prongs are developed in the last abdominal sternite. When the aphid is attacked the abdomen is bent and moved forward, the prongs are situated on the upper side and cover the aphid from above, thus preventing its escape. The ovipositor sheaths form the second part of the apparatus, which practically may be compared with vertically laid forceps.

In the genus *Metaphidius* the apparatus is of a different kind. It is formed by the tubiform prong at the base of tergite 7, which represents the recourse of the ovipositor sheaths in case of attack of the aphid, so that all the apparatus is based on a similar principle to that in *Trioxys*.

Anatomy of imago does not in general differ from that of other groups of *Hymenoptera* and for this reason it was omitted here. There are only differences in the internal female genitalia (described from *Diaeretiella rapae*): A pair of drop-like ovaria is developed, the shape being in dependence on their content. Terminal filaments are lacking. In the proximal portion of each ovary there is a mass of undistinguished cells, from which different groups of cells separate, and which are then moved to the oviduct. In this way the ovum becomes mature, the chorion being formed from the follicular epithel. The oviducts are united in a single, a little wider unpaired portion — the uterus, which narrows and opens into the vagina. On the dorsal side of the vagina, the receptaculum seminis opens, which is strongly sclerotized. Similarly, the unpaired opening of small spermathecal glands is sclerotized, too. On this side of the vagina there opens also the unpaired (? poisonous) sac. Besides, there opens into the vagina an unpaired opening of two big accessory glands that produce eosinophilic secretion which enables an egg to pass through the ovipositor. In the place of connection of these glands a spindle — form reservoir is developed.

From this scheme there are differences in the genera *Ephedrus* and *Praon*, where the ovariolae are prolongated as far as the base of the abdomen.

IV.

Taxonomy

Literary review

The aphidiids are an insect group that has long been known in spite of the small size of its representatives. The oldest information comes from the year 1695, in Leuwenhoek's Arcana Naturae.

The actual history of investigations on European aphidiids began with the 10th edition (1758) of Linné's Systema naturae, where *Ichneumon aphidum* was described. Unfortunately the type was lost, so that nowadays it cannot be decided what was understood by the name. The first general paper on the group was published in 1818 by Nees v. Esenbeck, and once again in a more comprehensive form in the year 1834. The generic classification used by Nees is obsolete now, but the author's good knowledge of the group, appearing in the identifying characters pointed out, is still evident. Nees limited himself mostly to forming "*sectiones*", often conformable to the modern conception of genus. Nees's collection was destroyed, and only exceptionally a species described by him can be redescribed. Nees's work was published a year later than Haliday's, let us say its first part, which caused confusion in the synonymy of individual species. Haliday (1833, 1834) is another original author of the group. His generic classification, of course extended and completed in the historical respect, has been used in principle up to the present. Haliday's collection, consisting mostly of material from the British Isles, was deposited in the British Museum, and most of the species described by him have been clarified. Wesmael (1835, see Starý, 1958) whose entire work on the *Braconidae* of Belgium is of basic importance though, was not very concerned with the aphidiids, and the species described by him have mostly been retracted as synonyms of the species described by Haliday. A few species, mostly parasites of forest aphid pests, were described by Ratzeburg (1844, 1848, 1852) in his monumental three-volumed work "Ichneumonen der Forstinsekten". The generic classification of the aphidiids has become clear and valuable only in the later volumes of the work. The fundamental generic classification of the aphidiids, rarely accepted by later authors and only nowadays properly appreciated, can be found in Förster's work (1862). Förster described many new genera. Unfortunately his diagnoses were very brief, differently understood by later authors, and since they had not been compared with types, many species were wrongly included in the individual genera and, moreover, have become deep-rooted in the applied entomology.

The first comprehensive picture of the European aphidiid fauna was given by Marshall (1896) in the 5th volume of André's "Species des Hymenoptères". In the year 1899 he published an abridged edition of the work, concerning the *Aphidiidae* of the British Isles. Marshall elaborated the specific composition of the *Aphidiidae* of Europe, which is the most positive feature of his work. The negative aspect is that he was describing males without knowing the females, so that most species were clarified only after the revision of types by Mackauer (1961). A certain step back was that Marshall had not used Förster's generic classification and considered it artificial in general. Nevertheless, his work is the best historical monograph on the group.

The *Aphidiidae* of Sweden were described by Thomson (1895). Apparently he limited himself to incomplete Haliday descriptions, and his identifications are not very reliable either. He did not make any essential changes in the generic classification of the group.

A not very critical catalogue of the aphidiids is in the 4th volume of the Catalogue of Hymenoptera by Dalla Torre (1898).

In 1900 the classification of European authors was used by Ashmead for the nearctic fauna, but not having known the types and working with incomplete original descriptions he made some mistakes.

A summary catalogue (within the family *Braconidae*) was published in the year 1904 by Szepligeti in the 22nd volume of Genera Insectorum. However, it was a mere list of well-known species.

Gahan (1911) revised the nearctic fauna of *Aphidiidae,* and his work meant a considerable progress in their classification.

Only in the thirties M. P. Quilis began to work on the taxonomy of *Aphidiidae*. He described several genera and species from southern and central Europe and a few fossil species and genera. The essential imperfection of his work was that for identification he used only historical literature without comparison with the types, and his knowledge of the range of variability of individual species was insufficient, so that most genera and species described by him have been retracted as synonyms after a revision of the types (see Starý, Mackauer).

Fahringer (1937) was concerned with *Aphidiidae* too, but most of his papers remained unpublished. His papers are not very valuable and precise, keys and descriptions made without the knowledge of types, often after descriptions only, and are without much value.

During World War II no important paper on aphidiids was published in Europe. At that time Watanabe was working on the fauna of Japan and Mongolia. In the year 1944 C. F. Smith published his outstanding work, a revision of the nearctic

Aphidiidae, which has become a basis of modern European compilations and revisions. It is a concise and clear work, with new taxonomic criteria.

Intensive investigations on the group, based mostly on the classification of biological species, began only in the year 1958, having been carried out up to the present with continuous confrontations and revisions of the author's opinions (Starý, Mackauer, 1. c.). European fauna has been investigated for the greatest part, and on this basis the research is extended to other areas.

The present research of the group is made in co-operation with authors from many countries, aphidologists as well as entomophagologists, and many a taxonomical, ecological, etc. problem is solved on a world-wide scale (Mackauer, Muesebeck, Schlinger, Starý, and others). On the basis of a roughly clarified generic and specific classification the parasites of aphids from many world areas are investigated, with particular regard to the employment of aphidiids in aphid pest control.

Czechoslovakia. Up to 1958 practically no data on the *Aphidiidae* from the territory of present day Czechoslovakia were available. The first information of the species *Praon dorsale* from the surroundings of Kaplice was given by Kirchner (1856). In Szepligeti's paper (1898) two finds of *Aphidiidae (Aphidius rosae, Praon flavinode)* from Slovakia were recorded. A few aphidiid species from the High Tatra were recorded by Niezabitowski (1909). Other data from the Czechoslovak Republic appeared only in the thirties. In 1934 Dragoun published his paper on the bionomy of *Lysiphlebus fabarum* (Marshall), parasite of *Aphis fabae* on sugar beet. In the same year Quilis M. P. published in Spain the first comprehensive work on the *Aphidiidae* from Czechoslovakia. A few new species were described there. His work was based on the material of *Aphidiidae* bred and sent him by Baudyš, who later published a short list of parasites reared by himself, with finding places of the hosts, in the year 1941 in Entomologické listy. In 1935 Schimitschek mentioned the find of *Pauesia pini* (Hal.), parasite of the aphid *Cinara laricis* Koch, in southern Moravia in the paper on the results of the breeding of parasites in Austria.

Further research of the *Aphidiidae* of Czechoslovakia began only in 1958 with the author's papeis. The investigations have been made within the European fauna of aphid parasites. When the principal taxonomic problems have been clarified, the ecological problems and recently also the biological control of aphids is considered an integral part of the complex problem of aphid control.

Taxonomic characters

The aphidiids are exclusively aphid parasites and, because of the relative homogeneity of the host group, their morphology is hardly different in many cases. The criteria of identification are of a different value and variability in individual genera and species. That is the reason they are presented here schematically only, being mostly included and figured in the key to genera and species.

Head. Shape transverse or square, strongly or arcuately narrowing behind eyes; the head is rarely wider across than it is behind the eyes. Dimensions on the head: There are some that are often good distinguishing characters: width of head in comparison with thorax, interocular line, facial line, transfacial line, clypeoantennal line, tentorio-ocular line, intertentorial line, socket-ocular line. Eyes are of various size and prominence, oval to nearly hemispherical, convergent to the clypeus. Clypeus is transverse, bearing long and more or less dense hairs. Antennae are filiform, rarely moniliform. The number of antennal segments is variable, mostly of sexual dimorphic character. The ratio of F_1 to F_2 is important. In some cases the number of rhinaria on F_1, F_2 and F_3 may be used as an additional character, but a series of specimens is necessary. The apical segment is of various shapes, sometimes fused with the praeapical segment.

Thorax. The declivity of mesoscutum to prothorax is sometimes a good character. The pubescence of the lobes of the mesoscutum is different — either there are rows of hairs along the (effaced) notaulices and margins, or hairless spots on the discs of lateral lobes. Notaulices are of a variable length and depth, sometimes entirely effaced; usually they are visible in the ascendant part of mesoscutum, being more or less rugose or crenulate and wide, and effaced on the disc; they rarely reach the praescutellar groove, being punctate or crenulate.

Wings. Shape, length and width of pterostigma. Length of metacarp. Length radial abscissa 1, relations between veins, etc. (Note: the variation in wing venation is sometimes remarkable and in some cases it may be of the "generic" value.)

Legs. Strength, length. Usually they are homogeneous, slender and long, rarely strong and short.

Abdomen. Shape. Tergite 1; it may be square, longer than wide; its sculpture is various — smooth, rugose, rugose-punctate, bearing carinae, etc.; primary (spiracular) or sometimes secondary tubercles are developed; the ratio of the width at spiracles to the length, and the ratio of the distance between spiracular tubercles and spiracular tubercle × distance from apex; width at spiracles × width at apex. External genitalia of females: shape, pubescence of ovipositor sheaths, to a lesser degree also the shape of prongs of valvulae 2, etc. Ovipositor sheaths may be triangular, trifid at the extremity; or prolongately arcuate, obtuse at the apex; or curved downwards, narrow or ploughshare-shaped; or curved downwards, widened at the apex, etc.

Coloration is sometimes a valuable additional character, sometimes it is variable. Most females are lighter than males. When classifying the coloration it is necessary to have material from various parts from the area of distribution, as in southern districts the coloration is lighter (the yellowish colour more distributed) than in northern areas.

27

Distribution. The same aphid species can be attacked by various parasites in different parts of the area. The parasites form faunistic complexes that are difficult to classify. Their distribution is contingent on the species and very inhomogeneous.

Habitat. *Aphidiidae* are comparatively stenotopic (see chapter VIII), so that the habitat is a good identifying character. Their partly mutable relation to the habitat within the area must be considered.

Host specificity. (See chapter VIII.) The relation to the systematic group of the host or its mode of life are of great importance for the host specificity of the parasite. In many cases of complete specificity of the parasite and host the individual parasitic species can be well distinguished in advance by their hosts. Sometimes the parasites specialize rather on the host's ecology — gall-producing, root and other aphids.

The manner of pupation and colour of mummified aphids may be a good additional character. Cocoons of *Praon* and *Dyscritulus* are a distinct character, distinguishing these genera from the other *Aphidiidae;* the aphids mummified by the *Ephedrus*-species are black.

Larval instars. Good generic characters can be found sometimes on the first instar larvae, which possess a well-developed cauda, or the cauda and two perpendicular prongs, or the cauda is long and arcuate and the prongs are oblique.

Specific criteria

Many criteria can be used in the classification of *Aphidiidae,* as the aphidiids parasitize exclusively aphids; in numerous material the influence of various factors on the development of individual groups of the parasites can be well observed. Naturally the specific criterion should be understood relative to the state of investigation into the group.

A. In the host

1. General knowledge of the group. A general knowledge of the phylogeny and classification of individual aphid groups and factors conducive to the settlement of new habitats, differentiation of the groups and their relationship is necessary. It helps in case of higher specialization of the parasite to understand its host specificity, requirements on the environment, etc., and to classify it.

2. Requirements on the environment. The requirements of aphids on the environment are various, and very important for the specific differentiation of the host; they often change in different parts of the area. The aphids include xerophilous, mesophilous and other forms; many aphid species change in connection with food

requirements the type of habitat and environment as a result of migration from the primary to secondary host plants (dioecious aphids). In connection with the specialization of the parasites, as we know it now, they thus belong in the range of various complexes of parasites which depend in a high degree on the type of habitat.

3. The distribution of aphids is in close connection with their host plants and requirements on the environment. Some aphid species are confined to a certain type of landscape (desert), others are conformable and nowadays distributed over the world as pests of cultures due to human economical activities; these become subject to various parasitic complexes in different areas.

4. Main morpho-ecological types. The aphids are largely differentiated ecologically. The ecological differentiation of aphids results in the morpho-ecological adaptation of the parasites. The aphids differentiate in gall-producing species, root aphids, species living in more or less dense colonies, aphids producing strong wax covers, etc. Sometimes the morpho-ecological type of an aphid species changes due to the peculiarities of the life-cycle — e.g. some aphids occur in forest habitats on primary host plats as gall producers, on secondary host plants in steppe type of habitats as root aphids, etc.

5. Bionomics. It is necessary to know the most important bionomical features of the main aphid groups. A complicated life cycle developed in many aphid species, being decisive for parasitization in different types of habitats.

6. Behaviour. The type of host behaviour is rather important. The parasites are also more or less adapted to the type of host behaviour. Some species of aphids are strongly movable, the others have strongly developed escape reactions and defense reactions, the others have saltatorial legs, etc.

B. In the parasite

1. General directions of phylogeny of higher taxonomic groups (superfamily, etc.). Aphids and parasites belong to two quite different insect orders — to the *Homoptera* and *Hymenoptera*. Each of these orders has its special developmental directions and for this reason it is incorrect to classify the relations between host and parasite without the knowledge of the above mentioned factors.

2. Main morpho-ecological types. A number of parasite genera and species represent cases of close adaptation of a parasite to host, which may be often classified as a convergence in evolution. These are typical morphological adaptations, enabling the more successful attack of the host aphid. In other cases, there are adapted parasites of gall aphids, leaf-curling aphids, root aphids, etc.

3. General rule influencing the host specificity. The host specificity of parasites is influenced by a number of factors (see chapter VII). To classify it, it is necessary to have a general outlook on the problem in other groups of entomophagous parasites.

4. Behaviour of parasites is specifical and very characteristic. The parasites are adapted to escape and defensive reactions of hosts, so that there is a different rate of oviposition, etc.

A classification of a species must be based on sufficient material from all the area of distribution. The material must be bred in a more or less big series. The bred material enables the ascertaining of the occurrence of males and females, etc. World knowledge of the group, at least in the main features, is also necessary as there are many widely distributed species, which are often described under different names. As for the description of a species, the following scheme is recommended for use: Differential diagnosis, descriptions of female and male, distribution, material, habitat, host-specificity, not to speak of generally accepted necessities such as figures, designation of type, etc.

A review of genera and species

All this chapter is based on the material of the Czechoslovak *Aphidiidae* exclusively. Only species ascertained in Czechoslovakia and known to the author are keyed. Similarly, all the records on the general distribution of individual species are also original; as to the literary data, references are given in individual genera. As in a number of cases the validity of some European species is unclear, the monographs and revision are recommended to be used in such cases, as the present book is not a monographic type.

The main task of this chapter is to show the relation of Czechoslovakia to other countries from the viewpoint of the distribution of *Aphidiidae*. On the basis of ecological data the greatest part of parasite species may be classified for the purpose of biological control of aphids.

In individual genera synonymy, distribution, habitat (with the emphasis on the type of habitat and foci), hosts (list of aphid hosts and aphid host plants; in case that hosts are unknown in Czechoslovakia, only a short note on literary data is presented to show the general feature of specificity of the given species), host specificity, occurrence (seasonal occurrence, occurrence in types of habitats, notes on the diapause, etc.), economic importance (indifferent or useful parasite), notes (more detailed data on ecology and bionomics, as far as they were studied in Czechoslovakia).

The genera are listed in accordance with their main phylogenetic relationship, further division on tribes and subfamilies has been omitted. The species are listed alphabetically, irrespective of subgenera.

30

Genus *Ephedrus* Haliday

Aphidius Nees subg. *Ephedrus* Haliday, 1833, Ent Mag. 1 : 261, 485
  Type-species: *Bracon plagiator* Nees
Elassus Wesmael, 1835. Nouv. Mém. Acad. Sci. Bruxelles 19 : 248
  Type: *Aphidius parcicornis* Nees

  Subgenera: *Ephedrus* s. str., Starý, 1958, Acta Faun. Ent. Mus. Nat. Pragae 3 : 66 – 67
  Type-species: *Ephedrus plagiator* (Nees)
Lysephedrus Starý, 1958, Acta Faun. Ent. Mus. Nat. Pragae 3 : 64
  Type-species: *Ephedrus validus* (Haliday)

  Revision: 1941, Stelfox, Proc. R. Irish Acad. 46 (B): 128 – 131. 1958, Starý, Acta Faun. Ent. Mus. Nat. Pragae 3 : 61 – 62; 1962, Starý, Opusc. entomol. 27: 87 – 98.

  Distribution: Pal., Nea., Aeth., Orient., Austr. reg.

  Note: Species of this genus have various, usually wider host-specificity. They parasitize a number of aphids except *Callaphididae, Lachnidae* and *Thelaxidae*.

1. *E. (E.) brevis* Stelfox

Ephedrus brevis Stelfox, 1941, Proc. R. Irish Acad. 46(B): 140 – 142

  Distribution: Br. Isles, Czechoslovakia.

  Habitat: This species was found only once in Czechoslovakia, in a *Corylus* and *Betula* wood in a wet habitat. A similar habitat is mentioned in the original description.

  Host: Unknown. Probably a certain aphid species living on *Corylus* or *Betula*.

  Host-specificity: Unknown.

  Occurrence: VIII.

  Economic importance: Unknown, probably none.

2. *E. (E.) campestris* Starý

Ephedrus (Ephedrus) campestris Starý, 1962, Opusc. entomol. 27: 87 – 91

  Distribution: Europe, Far East.

  Habitat: It occurs commonly in various kinds of steppe type habitats, everywhere where *Dactynotine* aphids may be found. It is rather numerous in steppe localities of the southern districts of Czechoslovakia, where *Centaurea stoebe* and other species are common. In cultivated landscape it occurs in balks in fields, roadsides, pathways and waste places, where *Artemisia, Achillea, Centaurea* or *Cichorium intybus* may be found, which plants are usually heavily infested by *Dactynotine* aphids.

  Host: *Dactynotus aeneus* HRL. on *Carduus nutans, Dactynotus cichorii* (Koch) on *Cichorium intybus, Leontodon hispidus, Dactynotus jaceae* (L) on *Centaurea stoebe,*

Centaurea scabiosa, Dactynotus muralis (Bckt.) on *Lactuca quercina, Dactynotus obscurus* (Koch) on *Hieracium* sp., *Dactynotus picridis* (F.) on *Picris hieracioides, Dactynotus sonchi* (L.) on *Sonchus oleraceus, Dactynotus* spp. on *Crepis biennis, Carlina vulgaris, Leucanthemum vulgare, Macrosiphoniella absinthii* (L.) on *Artemisia absinthium, Macrosiphoniella millefolii* (Deg.) on *Achillea millefolium, Achillea nobilis.*

Host-specificity: A parasite of *Dactynotus* and *Macrosiphoniella*-species.

Occurrence: VI, VII, VIII, IX.

Economic importance: Economically indifferent species because of its host-specifity. Sometimes it is rather effective, attacking heavily colonies of *Dactynotini.*

Notes: Mummified aphids are black.

3. E. (E.) cerasicola Starý

Ephedrus (Ephedrus) cerasicola Starý, 1962, Opusc. entomol. 27: 91—92

Distribution: Czechoslovakia.

Habitat: It occurs in forest type habitats — in parks and orchards.

Host: *Myzus cerasi* (F.) on *Prunus avium.*

Host-specificity: It has been known as a parasite of *Myzus cerasi* (F.) but it is believed it can parasitize some other aphid species too.

Occurrence: VI.

Economic importance: It occurs commonly with two other parasites of *Myzus cerasi* (F.) — *Ephedrus persicae* Froggatt and *Ephedrus plagiator* (Nees). The mentioned two species control the aphid to a certain degree, but the occurrence of diapause-cocoons in the first species lowers its effectiveness.

Note: Mummified aphids are black.

4. E. (E.) lacertosus (Haliday)

Aphidius (Ephedrus) lacertosus Haliday, 1833, Ent. Mag. 1: 486

Distribution: Europe (Br. Isles, Czechoslovakia).

Habitat: As far as it is known it occurs namely in forest undergrowth. Once it was found in a park too.

Host: *Macrosiphum rosae* (L.) on *Rosa* sp., *Rhopalosiphoninus* sp. on *Oxalis acetosella.*

Host-specificity: *Rhopalosiphoninus* sp. seems to be its main host.

Occurrence: V, VI, VII, VIII.

Economic importance: Economically indifferent species.

Note: Mummified aphids are black.

5. E. (E.) minor Stelfox

Ephedrus minor Stelfox, 1944, Ent. mon. Mag. 80: 235—236

Ephedrus plagiator (Nees) var. *minor* Stelfox, 1941, Proc. R. Irish
  Acad. 46 (B): 135–136

Distribution: Europe (Br. Isles, Czechoslovakia).

Habitat: It occurs in forest type habitats, to a lesser degree in intermediate habitats too. It may be found mostly in parks and gardens or in field shrubs in balks, roadsides, etc.

Host: *Myzaphis rosarum* (Kalt.) on *Rosa* spp., *Passerinia tetrarhoda* (Walk.) on *Rosa* spp.

Host-specificity: A parasite of *Myzaphidine*-aphids *(Myzaphis, Passerinia)*.

Occurrence: V, VI, VII.

Economic importance: Economically indifferent species. Sometimes it was found to be effective as parasite of *Myzaphis rosarum* (Kalt.).

Note: Mummified aphids are black.

6. E. (E.) nacheri Quilis M. P.

Ephedrus nacheri Quilis M. P., 1934, Eos Madrid 10: 17–19

Distribution: Europe (Czechoslovakia).

Habitat: It occurs in steppe type habitats, mostly in waste places, unsatisfactorily cultivated fields (weeds), roadsides, gardens.

Host: *Cryptosiphum artemisiae* Bckt. on *Artemisia vulgaris, Hayhurstia atriplicis* (L.) on *Chenopodium album, Chenopodium* spp.

Host-specificity: *Hayhurstia atriplicis* (L.) is its main host. The mode of host life is also important as it parasitizes also other aphids that cause galls on plants, whose taxonomic affinity is different.

Occurrence: VI, VII, VIII.

Economic importance: Because of its host -specificity it is an economically indifferent species. The percentage of infestation of *Hayhurstia* was found sometimes to be high.

Note: Mummified aphids are black.

7. E. (E.) persicae Froggatt

Ephedrus persicae Froggatt, 1904, Agric. Gaz. Sydney 15: 611–612
Ephedrus nevadensis Baker, 1909, Pomona J. Ent. 1: 23
Ephedrus nitidus Gahan, 1917, Proc. U.S. Nat. Mus. 53: No. 2197: 195
Ephedrus vidali Quilis M. P., 1931, Eos Madrid 7: 72–74
Ephedrus pulchellus Stelfox, 1941, Proc. R. Irish Acad. 46 (B): 133–139
Ephedrus interstitialis Watanabe, 1941, Ins. Matsum. 15: 139–140
Ephedrus impressus Granger, 1949, Mém. Inst. sci. Madagascar 2(A): 412

Ephedrus (Ephedrus) holmani Starý, 1958, Acta Faun. Ent. Mus. Nat. Pragae 3: 63, 68—70

Ephedrus (Ephedrus) palaestinensis Mackauer, 1959, Beitr. Ent. 9: 867—868

Distribution: Almost cosmopolitan.

Habitat: This species occurs rather commonly in forest type habitats as it may be recognized from a large number of samples we collected during 6 years in Czechoslovakia. It may be found namely in parks, orchards, gardens, vineyards, etc. where its host aphids usually occur most commonly too. Nevertheless, it may be found as a parasite of economically indifferent aphid hosts in clearings of woods, small woods in fields, avenues of *Sorbus* along submountain roads. Its occurrence in steppe type habitats is exceptional.

Host: *Allocotaphis quaestionis* (Börner) on *Malus silvestris*, *Aphis fabae* Scop. on *Euonymus europaea*, *Matricaria inodora*, *Brachycaudus helichrysi* (Kalt.) on *Persica vulgaris*, *Anthemis* sp., *Hieracium laevigatum*, *Matricaria suaveolens*, *Aphis idaei* v. d. G. on *Rubus idaeus*, *Brachycaudus lychnidis* (L.) on *Melandrium* sp., *Silene cucubalus*, *Brachycaudus* sp. on *Prunus domestica*, *Prunus spinosa*, *Melandrium* sp., *Dysaphis devecta* (Walk.) on *Malus silvestris*, *Dysaphis sorbi* (Kalt.) on *Sorbus aucuparia*, *Dysaphis* spp. on *Malus silvestris*, *Pirus communis*, *Sorbus torminalis*, *Hyadaphis mellifera* (Hottes) on *Lonicera xylosteum*, *Myzodes ligustri* (Mosl.) on *Ligustrum aviculare*, *Myzus cerasi* (F.) on *Prunus avium*, *P. cerasus*, *Phorodon humuli* Schrk. on *Prunus domestica*, *Humulus lupulus*, *Rhopalosiphum padi* (L.) on *Padus racemosa*, *Roepkea marchali* (Börner) on *Prunus mahaleb*.

Host-specificity: It is a common parasite of various leaf-curling aphids. Although it is clear the *Anuraphidine* group is probably a source of its phylogenetical hosts, it attacks today a number of aphids connected with *Rosaceae*. Nevertheless, from the comparison of percentage of its occurrence in various habitats, on various hosts in dependence on the mode of their life, it is apparent that today it may be classified as a typical parasite of leaf -curling aphids, of namely *Anuraphidine* and *Myzine* groups, which occur in forest type habitats. In the mentioned aphid groups the main hosts of the parasite belong. As subsidiary hosts *Rhopalosiphum padi* L., *Aphis idaei* v. d. G. may be mentioned. The parasitization of *Aphis fabae* Scop. may be classified as almost exceptional. As a certain kind of criterion for the differentiation of main, subsidiary and facultative hosts the presence or absence of diapause cocoons may be also used. It is probable that the parasite, i. e. its diapause stage, needs a more suitable host; the progeny that was laid in a less suitable host probably would not survive the diapause period. For this reason the diapause cocoons seem to occur in the main (subsidiary) hosts only.

Occurrence: V, VI, VII.

Economic importance: It is a typical parasite of leaf-curling aphids, which often include serious pests in orchards (*Myzus cerasi* F., *Dysaphis* spp., etc.). From the comparatively high fecundity and the high percentage of aphid infestation, which was ascertained in the field, we may conclude that this species could be a relatively effective parasite. However, its effectiveness is stopped very soon in the spring as the greatest part of specimens fall in diapause in this period. At this time, however, the host is still very common and an outbreak could occur under certain conditions. It is possible that diapause cocoons are in certain parts of the total distribution area, while they are not formed in other districts, where the species could then reach high effectiveness. However, the contemporary state of our knowledge of the distribution and ecology of this species in other areas is too poor to give any definite answer to this question.

Note: First instar larva has 13 segments and the head. The last segment is caudate, bearing two perpendicular processess at its base. The dorsal part of the cauda is covered by small spines. There is a narrow transverse row of spines in each segment, similarly as in the first instar larva of *Ephedrus plagiator* (Nees).

Oviposition behaviour is similar to that of other *Aphidiidae*. The oviposition lasts about 10−15 seconds, corresponding to the slow escape reactions of the host aphid *Myzus cerasi* F. Under suitable conditions the female attacks are made comparatively systematically, but it often runs a little so that the parasitized aphids occur separately, never in dense groups. The attacked aphid tries to repel the attacking female parasite by movements of its legs or it runs off; these movements however, do not influence the attacking female, as in some other aphidiid species; it bends the abdomen appropriately or follows the running aphid, but the ovipositor remains inserted.

Fecundity of females was studied by the dissection of emerged virgin females reared for several hours without food. About 70 developed eggs and a big quantity of undeveloped eggs were found in each ovary, so that the species is a synovigenous parasite.

Judging from the series of material collected in the field and reared in the laboratory, which include both males and females, it is the arrhenotokous type of parthenogenesis that is developed in this species in Czechoslovakia.

Diapause. During several years two types of cocoons, i.e. of mummified aphids were observed in *Ephedrus persicae* Frog. to occur in Czechoslovakia. Further studies have shown that they represent diapause and non-diapause cocoons.

a) Non-diapause cocoons. The mummified aphids have the usual appearance of dead parasitized − mummified aphids. They are mat, smaller in relation to diapause cocoons.

b) Diapause cocoons. The mummified aphids are unusually large, globular in shape, shiny, strongly mummified, the segmentation of the abdomen being unrecognizable on them. They have the appearance of dark shiny balls.

The percentage of the occurrence of diapause and non-diapause cocoons is rather variable. It was very high usually, reaching sometimes nearly 100%, while in other cases mostly or exclusively non-diapause cocoons occurred.

Dissections of diapause cocoons have shown that the diapause is spent in the mature larva stage. As our laboratory observations have shown the diapause in this species lasts all (or the rest of) Summer and Autumn, and it is finished after over-wintering, when after a certain period the mature larva pupates and imago then emerges.

Cocoons. Mummified aphids (non-diapause cocoons) are black, diapause-cocoons are brownish.

Augmentation in orchards. According to our studies, the augmentation of parasites by using certain indifferent aphid species as alternative hosts has no reason to be undertaken. It might be possible to use *Dysaphis sorbi* (Kalt.) as an economically indifferent aphid host to augment *Ephedrus persicae* Froggatt in the neighbouring orchards, but the diapause-cocoons occur in all the main hosts, both economic pests and indifferent species, so that because of the diapause the species cannot be augmented by this way.

8. E. (E.) plagiator (Nees)

Bracon plagiator Nees, 1811, Mag. Ges. naturf. Fr. Berlin 5: 17
Aphidius parcicornis Nees, 1834, Hym. Ichn. aff. Mon. 1: 16
Ephedrus japonicus Ashmead, 1906, Proc. U. S. Nat. Mus. 30: 187
? *Ephedrus plagiator* (Nees) var. *nigra* Gautier, Bonnamour, Gaumont, 1929, Bull. Soc. ent. Fr. 1929: 200–201

Distribution: Pal. reg.

Habitat: The species is comparatively widely eurytopic. Its main habitats are without any doubt habitats of the forest type. There it may be found in undergrowths of deciduous and mixed forests, namely, in borders of woods and in clearings, in bushes in fields, in orchards, avenues of shady trees, etc. To a certain degree it attacks also some aphids (*Sitobium, Rhopalosiphum,* on grasses) in fields i.e. in steppe habitats too. In the submountain districts where it is common in corn fields, this might be caused by the similar character of small fields which are situated among woods and could be classified as habitats of an intermediate type. In lowlands, this is but a mark of its wider ecological plasticity that is in connection with the existence of preferred aphid hosts *(Sitobium, Rhopalosiphum)* in the field habitats too. Nevertheless, this adaptation for the occurrence in steppe habitats is apparently secondary as it may be recognized from a large number of samples we collected in Czechoslovakia, from which it is clear that forest type habitats are clearly preferred.

Host: *Acyrthosiphon caraganae* Chol. on *Caragana arborescens, Acyrthosiphon spartii* (Koch) on *Sarothamnus scoparius, Aphis craccae* (L.) on *Vicia cracca, Aphis fabae* Scop. on *Euonymus europaea, Borago officinalis, Philadelphus coronarius, Impatiens roepkei, Cirsium* sp., *Beta vulgaris, Chenopodium* sp., *Epipactis latifolia, Valeriana officinalis, Aphis farinosa* Gmel. on *Salix* spp., *Aphis idaei* v. d. G. on *Rubus idaeus, Aphis nasturtii* (Kalt.) on *Rhamnus cathartica, Aphis pomi* Deg. on

Crataegus monogyna, Aphis spiraephaga Müller on *Spiraea* spp., *Aphis urticata* (F.) on *Urtica urens, Urtica dioica, Aphis* spp. on *Robinia pseudoacacia, Aulacorthum chelidonii* (Kalt.) on *Chelidonium majus, Aulacorthum* spp. on *Naumburgia thyrsiflora, Potentilla argentea, Brachycaudus cardui* (L.) on *Prunus spinosa, Prunus cerasifera, Brachycaudus* spp. on *Prunus domestica, Prunus* sp., *Ceruraphis eriophori* (Walk.) on *Viburnum opulus, Chaitophorus* sp. on *Populus* sp., *Dysaphis devecta* (Walk.) on *Malus silvestris, Dysaphis sorbi* (Kalt.) on *Sorbus aucuparia, Dysaphis* spp. on *Pirus communis, Sorbus torminalis, Malus silvestris, Crataegus* sp., *Hyalopterus pruni* (Geoffr.) on *Prunus domestica, Hyperomyzus lactucae* (L.) on *Ribes nigra, Ribes grossularia, Liosomaphis berberidis* (Kalt.) on *Berberis vulgaris, Macrosiphum prenanthidis* Börner on *Prenanthes purpurea, Macrosiphum rosae* (L.) on *Rosa* sp., *Myzocallis coryli* (Goetze) on *Corylus avellana, Myzus cerasi* (F.) on *Prunus avium, Phorodon humuli* (Schrk.) on *Prunus domestica, Prociphilus fraxini* (Htg.) on *Fraxinus excelsior, Rhopalosiphum padi* (L.) on *Padus racemosa, Secale cereale, Schizaphis scirpi* Kittel on *Typha angustifolia, Schizoneura ulmi* (L.) on *Ulmus laevis, Sitobium avenae* (Fabr.) on *Rosa* sp., *Secale cereale, Sitobium equiseti* Holman on *Equisetum silvaticum, Sitobium* spp. on *Dactylis glomerata, Hordeum distichon, Avena sativa, Triticum vulgare, Secale cereale.*

Host-specificity: This is a widely specialized species. As the main host *Aphis, Brachycaudus, Ceruraphis, Dysaphis, Myzus, Hyalopterus, Hyperomyzus, Macrosiphum, Rhopalosiphum, Sitobium*-species may be mentioned. Generally, it mainly attacks aphid species that live in dense colonies or leaf-curling species.

As its phylogenetic hosts, aphids of the family *Aphididae* may be given, with a life-history of which the parasites in forest type habitats are connected. Gradually, because of its progressive character — wide specialization, it attacked also other aphid groups of a similar mode of life, occurring in the same type of habitats, so that the list of its hosts is rather numerous today.

Occurrence: V, VI, VII, VIII, IX, X.

Economic importance: This is economically a rather valuable species. It attacks a number of pest aphids in forest type habitats, namely the dioecious species, so that their number that migrate to fields on cultivated crops is lowered due to the parasite activity in forest habitats. Moreover, it is widely specialized, so that it may occur in the same type of habitats and parasitize a number of other aphid species. It may occur in a number of various habitats too. Diapause was also not observed to exist in Czechoslovakia. Activities on the augmentation might be done by using the indifferent aphidofauna in the neighbouring habitats of cultivated areas. Nevertheless, laboratory observations on its effectiveness must be undertaken as, judging from our field observation, it seems that the parasite fecundity is not too high.

Note: Mummified aphids are black.

The species was successfully reared in laboratory conditions (18 hours day light,

18−24°C, about 70% r. hum.) under neon light, *Aphis fabae* Scop. on *Vicia faba* plants being used as its host.

9. E. (L.) validus (Haliday)

Aphidius (Ephedrus) validus Haliday, 1833, Ent. Mag. 1: 485

Distribution: Europe (Br. Isles, Czechoslovakia), Far East.

Habitat: It was found by using the sweeping method to occur in forest type habitats (undergrowth, clearings).

Host: Unknown in Czechoslovakia. It is known from the Far East as a parasite of *Eriosoma*-species.

Host-specificity: Unknown, probably a parasite of leaf-curling aphids, or aphids that produce strong wax covers, etc. (pubescent ovipositor sheaths).

Occurrence: VIII.

Economic importance: Unknown in Czechoslovakia. Because of its parasitization of *Eriosoma*-species in the other areas it might be valuable. Nevertheless, it seems to be rare in Czechoslovakia.

Genus *Toxares* Haliday

Trionyx Haliday, 1833, Ent. Mag. 1: 487 (preocc.)
  Type-species: *Aphidius (Trionyx) deltiger* Haliday
Toxares Haliday, 1840, in Westwood, 1840, Introd. Mod. Class. Insects 2, Synopsis, p. 65
  Type-species: *Aphidius (Trionyx) deltiger* Haliday

Revision: 1941, Stelfox, Proc. R. Irish Acad. Dublin 46(B): 127−128. 1958, Starý, Acta Faun. Ent. Mus. Nat. Pragae 3: 58−59.

Distribution: Europe.

1. T. deltiger (Haliday)

Aphidius (Trionyx) deltiger Haliday, 1833, Ent. Mag. 1: 487
Ephedrus flaveolus Györfi, 1958, Acta Zool. Hung. 4: 131

Habitat: It occurs in forest type habitats, in deciduous forests and parks.

Host: *Acyrthosiphon caraganae* Chol. on *Caragana arborescens*.

Host-specificity: It is known as a parasite of a number in deciduous forests, orchards, and parks, living aphids that in many cases have no taxonomic affinity (*Drepanosiphum, Dysaphis, Acyrthosiphon, Myzus*, etc.).

Occurrence: VI.

Economic importance: Economically indifferent species.

Genus *Areopraon* Mackauer

Areopraon Mackauer, 1959, Beitr. Ent. 9: 849 – 850

  Type-species: *Praon lepelleyi* Waterston

  Revision: Mackauer, 1959, Beitr. Ent. 9: 849 – 854

  Distribution: Europe.

  Note: Species of this genus are probably specialized parasites of gall-producing aphids *(Schizoneura)*. Although this genus is closely related to *Praon* Haliday, its representatives spine the cocoon inside mummified aphid skin, while in the genus *Praon* Hal. a separate cocoon is spun under the empty aphid skin.

1. *A. lepelleyi* (Waterston)

Praon lepelleyi Waterston, 1926, Ent. mon. Mag., London 62: 237

  Distribution: Europe (Czechoslovakia).

  Habitat: This is a typical inhabitant of deciduous woods, parks, gardens.

  Host: *Schizoneura ulmi* (L.) on *Ulmus campestris*.

  Host-specificity: A typical parasite of *Schizoneura*-aphids forming galls on leaves of primary host plants *(Ulmus)*.

  Occurrence: V, VI, VII.

  Economic importance: It was found to be an effective parasite of *Schizoneura*-aphids on *Ulmus* in some cases.

Genus *Praon* Haliday

Aphidius Nees subg. *Praon* Haliday, 1833, Ent. Mag. 1: 483

  Type-species: *Bracon exoletus* Nees
Aphidaria Provancher, 1886, Add. Faun. Canad. Hym., p. 151

  Type-species: *Aphidaria simulans* Provancher

  Revision: Mackauer, 1959, Beitr. Ent. 9: 818 – 862

  Distribution: Pal., Nea.

  Note: Species of this genus spin a separate cocoon under the empty aphid skin. Instar I larvae possess 2 perpendicular appendages, besides the cauda.

1. *P. abjectum* (Haliday)

Aphidius (Praon) abjectus Haliday, 1833, Ent. Mag. 1: 485

  Distribution: Europe.

  Habitat: This is a common species in deciduous and mixed woods, parks, orchards, thickets and shrubs.

Hosts: *Aphis bupleuri* (Börner) on *Bupleurum falcatum*, *Aphis craccivora* (Koch) on *Robinia pseudoacacia*, *Aphis fabae* Scop. on *Philadelphus coronarius*, *Cirsium rigens*, *Arctium* sp., *Euonymus europaea*, *Aphis farinosa* Gmel. on *Salix* spp., *Aphis sambuci* (L.) on *Sambucus nigra*, *Aphis viburni* Scop. on *Viburnum opulus*, *Aphis* spp. on *Epilobium parviflorum*, *Epilobium montanum Rhopalosiphum padi* (L.) on *Padus racemosa*, *Rhopalosiphum* sp. on *Amygdalus nana*.

Host-specificity: A typical parasite of *Aphis* -species and allied genera (*Rhopalosiphum* spp.) in forest type habitats.

Occurrence: V, VI, VII.

Economic importance: This is a rather effective species. Heavy infestation of *Aphis fabae* Scop. on *Euonymus europaea* was often observed in various districts in Czechoslovakia.

Notes: This species was successfully bred in laboratory conditions under fluorescent light (18 hours day light) on *Aphis fabae* Scop., *Vicia faba* being laboratory host plant.

The parasite female is able to oviposit also in aphids that are situated between two closely related leaves. The main orientation is made by using the antennae, the secondary orientation is made by the ovipositor. The sting lasts about 2–3 seconds.

The oviposition is almost continuous, only comparatively short resting periods were observed. Higher aphid instars and adult aphids were clearly preferred, the macro-orientation being supposed to be most important in this case. Dense groups of lower (I–II) instar aphids were ignored or poorly attacked. Such groups of aphids were never systematically parasitized as e.g. in *Lysiphlebus fabarum* (Marsh.). Nevertheless, groups of higher instar aphids were observed to be parasitized in a similar way as in the mentioned parasite species. The mentioned preference of adult aphids by the ovipositing female seems to have a big importance in the passive spread of the parasite.

The adult parasites were observed to fly very often, searching for aphids. In nature, they were often observed to be active in aphid colonies on quiet sunny spring days, at about 4–5 hours p. m.

Aphid responses to the parasite female attack were ignored by the female, the stick being uninterrupted during aphid movements.

2. *P. absinthii* Bignell

Praon absinthii Bignell, 1894, Ent. mon. Mag. 30: 255–256

Distribution: Europe.

Habitat: This is a typical inhabitant of various steppe type habitats, occurring commonly in meadows, waste places, steppe preservations, ditches, etc.

40

Host: *Macrosiphoniella absinthii* (L.) on *Artemisia absinthium, Macrosiphoniella millefolii* (Deg.) on *Achillea millefolium, Achillea nobilis, Titanosiphon artemisiae* (Koch) on *Artemisia campestris.*

Host-specificity: This is a typical parasite of *Macrosiphoniella*-species.

Occurrence: VI, VII, VIII.

Economic importance: Economically indifferent species.

3. *P. bicolor* Mackauer

Praon bicolor Mackauer, 1959, Beitr. Ent. 9: 824–825

Distribution: Europe.

Habitat: Coniferous (pine) forests, mixed woods.

Host: *Protolachnus agilis* (Kalt.) on *Pinus silvestris.*

Host-specificity: A specialized parasite of *Protolachnus*-species.

Occurrence: VIII.

Economic importance: Economically indifferent species.

Note: The cocoons are whitish. The effectiveness seems to be small as parasite cocoons were observed to be very spread on a pine tree.

4. *P. dorsale* (Haliday)

Aphidius (Praon) dorsalis Haliday, 1833, Ent. Mag. 1: 484
Blacus discolor Nees, 1834, Hym. Ichn. aff. Mon. 1: 192
Praon collaris Förster, (nomen nudum, see Mackauer, 1959, Beitr. Ent. 9: 856)
Praon longicorne Marshall, 1896, in André, Spec. Hym. Eur. d'Alg. 5: 536–537

Distribution: Europe.

Habitat: It occurs in steppe type habitats, in meadows, waste places, alfalfa and clover fields, etc.

Host: *Acyrthosiphon pisum* (Harris) on *Vicia cracca, Trifolium pratense, Medicago sativa, Dactynotus campanulae* (Kalt.) on *Campanula rotundifolia, Dactynotus cichorii* (Koch) on *Cichorium intybus, Crepis biennis, Lapsana communis, Dactynotus jaceae* (L.) on *Centaurea stoebe, Centaurea scabiosa, Dactynotus linariae* (Koch) on *Aster linosyris, Dactynotus taraxaci* (Kalt.) on *Taraxacum officinale, Paczoskia major* Börner on *Echinops sphaerocephalus, Megoura viciae* Bckt. on *Lathyrus silvester, Lathyrus* sp.

Host-specificity: A typical parasite of *Dactynotus*-species, to a lesser degree it attacks also other aphid genera *(Acyrthosiphon, Megoura)* in steppe type habitats.

Occurrence: VI, VII, VIII, IX, X.

Economic importance: Being a parasite of *Dactynotus*-species, it is an economically indifferent species. As for its parasitization of *Acyrthosiphon pisum* (Harris), its effectiveness was found to be low in Czechoslovakia in comparison with that of *Aphidius ervi* Hal., the main parasite of this aphid.

Note: The cocoons are yellowish or whitish or brownish, usually in dependence on the aphid species attacked.

5. *P. exoletum* (Nees)

Bracon exoletus Nees, 1811, Mag. Ges. naturf. Fr. Berlin (1811): 30
Praon palitans Muesebeck, 1956, Bull. Brooklyn ent. Soc. 51: 27—28

Distribution: Europe, N. Africa, C. Asia, introduced into Nea.

Habitat: This is a typical inhabitant of steppe habitats. It occurs commonly in alfalfa fields; in roadsides, pathways, waste places, where *Melilotus* may be found. Such places represent probably its foci in nature.

Host-specificity: A specialized parasite of *Therioaphis species*.

Occurrence: V, VI, VII.

Economic importance: This is a very effective parasite of *Therioaphis*-species infesting alfalfa, being rather common in southern districts.

Note: The cocoons may be found on the upper side of leaves being whitish in colour.

6. *P. flavinode* (Haliday)

Aphidius (Praon) flavinodis Haliday, 1833, Ent. Mag. 1: 485

Distribution: Europe.

Habitat: Deciduous woods, parks, gardens, avenues.

Host: *Eucallipterus tiliae* (L.) on *Tilia tomentosa, Tilia europaea, Tilia cordata, Myzocallis carpini* (Koch) on *Carpinus betulus, Tinocallis platani* (Kalt.) on *Ulmus* sp., *Tuberculoides annulatus* (Htg.) on *Quercus* spp.

Host-specificity: This is a common parasite of a number of Callaphidid aphids occurring in deciduous woods and similar habitats of forest type.

Occurrence: V, VI, VII.

Economic importance: It is an effective parasite of the above mentioned aphids, which are however of no great economic importance in Czechoslovakia, except, maybe, of *Eucallipterus tiliae* (L.) which attacks *Tilia*-trees and on its honey-dew various fungi often develop.

Note: The cocoons are whitish. Mostly alate adult aphids were found to be placed on the top of cocoons.

7. *P. necans* Mackauer

Praon necans Mackauer, 1959, Beitr. Ent. 9: 839—841

Distribution: Europe.

42

Habitat: It was collected in a forest and intermediate habitats, in gardens, parks, etc., everywhere where small water reservoirs covered with various plants may be found. It may be found commonly in peat-bogs, ponds, etc., too.

Host: *Rhopalosiphum nymphaeae* (L.) on *Menianthes trifoliata, Hydrochaeris morsus-ranae, Nymphaea* sp.

Host-specificity: This is probably a rather specialized parasite of *Rhopalosiphum nymphaeae* (L.).

Occurrence: VII, VIII.

Economic importance: Due to its specificity it might be an economically significant species.

Note: Cocoons are whitish.

8. *P. pubescens* Starý

Praon pubescens Starý, 1961, Acta Soc. ent. Čechoslov. 58: 340—341

Distribution: Europe (Czechoslovakia).

Habitat: This is a common species occurring in undergrowth of woods, parks.

Host: *Nasonovia nigra* HRL. on *Hieracium silvaticum, Nasonovia ribisnigri* (Mosl.) on *Hieracium* sp.

Host-specificity: A strictly specialized parasite of *Nasonovia*-species.

Economic importance: This is a common and comparatively effective parasite, a member of the typical parasite complex of *Nasonovia*-aphids. As some species of this aphid genus are pest aphids, the parasite might be economically important under certain circumstances (in gardens on lettuce, etc.).

Note: Cocoons are whitish.

9. *P. rosaecola* Starý

Praon rosaecola Starý, 1961, Acta Soc. ent. Čechoslov. 58: 341—343

Distribution: Europe (Czechoslovakia).

Habitat: Forest and intermediate habitats — gardens, deciduous and mixed woods, parks.

Host: *Macrosiphum rosae* (L.) on *Rosa* spp.

Host-specificity: A parasite of *Macrosiphum rosae* (L.).

Occurrence: VII.

Economic importance: Economically indifferent species.

Note: Cocoons are whitish.

10. *P. volucre* (Haliday)

Aphidius (Praon) volucris Haliday, 1833, Ent. Mag. 1: 484

Distribution: Europe, Caucasus, Transcaucasia.

Habitat: It occurs mostly in a forest type and intermediate habitats but may be found in steppe habitats, too, being probably a comparatively eurytopic species. Nevertheless, its occurrence in forest type habitats is more typical judging from the material collected.

Host: *Acyrthosiphon caraganae* (Chol.) on *Caragana arborescens*, *Aphis craccivora* (Koch) on *Caragana arborescens*, *Brachycaudus helichrysi* (Kalt.) on *Melandrium album*, *Brachycaudus lychnidis* (L.) on *Melandrium rubrum*, *Melandrium album*, *Brevicoryne brassicae* (L.) on *Brassica oleracea* var. *capitata*, *Dysaphis* sp. on *Malus silvestris*, *Hyalopterus pruni* (Geoffr.) on *Prunus domestica*, *Prunus spinosa*, *Prunus armeniaca*, *Phragmites communis*, *Hyperomyzus lactucae* (L.) on *Sonchus oleraceus*, *Macrosiphum euphorbiae* Theo. on *Euphorbia cyparissias*, *Macrosiphum rosae* (L.) on *Rosa* sp., *Microlophium evansi* (Theo.) on *Urtica dioica*, *Sitobium* spp. on *Avena sativa*.

Host-specificity: This is a widely specialized species, parasitizing a number of various aphid groups.

Occurrence: V, VI, VII, VIII.

Economic importance: This is a rather effective species, in many cases a parasite of pest aphids too. Because of its wide specificity it might be of economic importance.

Note: Cocoons are whitish.

Genus *Dyscritulus* Hincks

Dyscritus Marshall, 1896, in André Spec. Hym. Eur. d'Alg. 5: 532, 613 (Preocc.)
   Type-species: *Dyscritus planiceps* Marshall
Dyscritulus Hincks, 1943, Entomologist, London 76: 104, 224

   Type-species: *Dyscritus planiceps* Marshall

   Revision: Starý, 1949, Acta Faun. Ent. Mus. Nat. Pragae 5: 69−74

   Distribution: Europe.

   Note: The species spin a separate cocoon under the empty aphid skin, its borders being more strongly spun, differring clearly from the rest of cocoon.

1. *D. planiceps* (Marshall)

Dyscritus planiceps Marshall, 1896, in André, Spec. Hym. Eur. d'Alg. 5: 618

   Distribution: Europe.

   Habitat: Deciduous and mixed woods, parks, avenues.

   Host: *Drepanosiphon platanoidis* (Schrk.) on *Acer platanoides*, *Acer pseudoplatanus*.

   Host-specificity: A strictly specialized species.

44

Occurrence: VI, VII, VIII.

Economic importance: Economically indifferent species.

Notes: Cocoons are yellowish brown. Only alate adult aphids were found to be placed on the top of cocoon.

Genus *Pauesia* Quilis M. P.

Pauesia Quilis M. P., 1931, Eos Madrid 7: 67–69
  Type-species: *Pauesia albuferensis* Quilis M. P.
Aphidius Nees, subg. *Paraphidius* Starý, 1958, Acta Faun. Ent. Mus. Nat. Pragae 3: 56, 91

  Revision: Starý, 1960, Acta Faun. Ent. Mus. Nat. Pragae 6: 5–44

  Distribution: Pal., Nea., Orient., Ethiop.

  Note: Pupation inside mummified aphid. Specialized parasites of Lachnid aphids.

1. *P. abietis* (Marshall)

Aphidius abietis Marshall, 1896, in André, Spec. Hym. Eur. d'Alg. 5: 565–566

  Distribution: Europe.

  Habitat: Coniferous and mixed woods.

  Host: *Cinara laricis* (Walk.) on *Larix europaea*, *Cinara* sp. on *Larix europaea*.

  Host-specificity: A parasite of *Cinara-species*.

  Occurrence: VI, VII, VIII.

Economic importance: Because of the little importance of its host this parasite may be classified as economically indifferent.

  Note: Mummified aphids are black.

2. *P. cupressobii* (Starý)

Paraphidius cupressobii Starý, 1960, Acta Faun. Ent. Mus. Nat. Pragae 6: 11–13

  Distribution: Europe (Czechoslovakia).

  Habitat: Submountain meadows, meadow-pastures, with *Juniperus*-shrubs.

  Host: *Cupressobium juniperi* (Deg.) on *Juniperus communis*.

  Host-specificity: This is probably a specialized parasite of *Cupressobium*-species.

  Occurrence: V, VI, VII.

  Economic importance: Economically indifferent species.

  Note: Mummified aphids are greyish black.

3. *P. grossa* (Fahr.)

Coelonotus grossus Fahringer, 1937, Festschr. 60. Geb. E. Strand, Riga 3: 244–245

Distribution: Europe (Czechoslovakia, Austria).

Habitat: Coniferous (fir) forests.

Host: *Todolachnus abieticola* Chol. on *Abies alba*.

Host-specificity: This is probably a specialized parasite of *Todolachnus abieticola* Chol.

Occurrence: VIII.

Economic importance: A very effective species. In case of the host's outbreak it might be economically important.

Note: Mummified aphids are black.

4. *P. infulata* (Haliday)

Aphidius infulatus Haliday, 1834, Ent. Mag. 2: 96
Paraphidius albiflagellaris Starý, 1960, Acta Faun. Ent. Mus. Nat. Pragae 6: 10–11

Distribution: Europe, Far East.

Habitat: Coniferous and mixed woods.

Host: *Buchneria pectinatae* (Nördl.) on *Abies alba*.

Host-specificity: A parasite of *Cinara-* (and *Cupressobium-*) species.

Occurrence: VI, VII.

Economic importance: In case of *Buchneria*-outbreak it might have a certain economic importance.

Note: Mummified aphids are yellowish-brown or blackish.

5. *P. juniperorum* (Starý)

Paraphidius juniperorum Starý, 1960, Acta Faun. Ent. Mus. Nat. Pragae 6: 16–17

Distribution: Europe (Czechoslovakia).

Habitat: Submountain meadows and pasture-meadows with *Juniperus*-shrubs.

Host: *Cupressobium juniperi* (Deg.) on *Juniperus communis*.

Host-specificity: This is probably a specialized parasite of *Cupressobium*-species.

Occurrence: VI, VII.

Economic importance: Economically indifferent species.

Note: Mummified aphids are blackish.

6. *P. laricis* (Haliday)

Aphidius laricis Haliday, 1834, Ent. Mag. 2: 97

Distribution: Europe.

Habitat: Coniferous woods, mixed woods.

Host: *Cinara* ssp. on *Pinus silvestris, Picea excelsa.*

Host-specificity: A parasite of *Cinara*-species.

Occurrence: V, VI, VII.

Economic importance: Economically indifferent species.

Note: Mummified aphids are brownish or blackish.

7. *P. maculolachni* (Starý)

Paraphidius maculolachni Starý, 1960, Acta Faun. Ent. Mus. Nat. Pragae 6: 19—20

Distribution: Europe (Czechoslovakia).

Habitat: As far as it is known this species occurs in parks, shrubs, where *Rosa* spp. occur.

Host: *Maculolachnus submacula* (Walk.) on *Rosa* sp.

Host-specificity: A specialized parasite of *Maculolachnus submacula* (Walk.).

Occurrence: Detailed data unknown.

Economic importance: Economically indifferent species.

8. *P. piceaecollis* (Starý)

Paraphidius piceaecollis Starý, 1960, Acta Faun. Ent. Mus. Nat. Pragae 6: 20—22

Distribution: Europe (Czechoslovakia).

Habitat: Coniferous woods.

Host: *Cinara* spp. on *Picea excelsa.*

Host-specificity: A parasite of *Cinara*-species, probably those occurring on spruce.

Occurrence: VII, VIII.

Economic importance: Economically indifferent species.

Note: Mummified aphids are blackish.

9. *P. pini* (Haliday)

Aphidius pini Haliday, 1834, Ent. Mag. 2: 96

Distribution: Europe, Far East.

Habitat: Coniferous woods.

Host: *Cinara* spp. on *Pinus silvestris, Larix europaea.*

Host-specificity: A parasite of *Cinara*-species.

Occurrence: VI, VII, VIII.

Economic importance: Economically indifferent species. In case of its host's outbreak it might be useful.

Note: Mummified aphids are brownish or blackish. They are usually very spread on an aphid host plant, being situated at the tops of pine- needles.

10. *P. silvestris* (Starý)

Paraphidius silvestris Starý, 1960, Acta Faun. Ent. Mus. Nat. Pragae 6: 29 – 31

Distribution: Europe (Czechoslovakia).

Habitat: Coniferous woods.

Host: *Cinara* spp. on *Pinus silvestris*.

Host-specificity: A parasite of *Cinara*-species.

Occurrence: VI, VII.

Economic importance: Economically indifferent species.

Note: Mummified aphids are blackish or brownish.

11. *P. unilachni* (Gahan)

Aphidius unilachni Gahan, 1926, Proc. U.S. Nat. Mus. 70; 8: 1 – 2
Pauesia albuferensis Quilis M. P., 1931, Eos Madrid 7: 69 – 71
Aphidius praevisus Gautier and Bonnamour, 1936, Bull. Soc. linn. Lyon N. S. 5: 74 – 75
Trioxys basilewskyi Benoit, 1955, Ann. Mus. Congo Belge (Sér. 8) Sci. zool. 36: 347

Distribution: Europe, Far East, Ethiopian reg.

Habitat: Coniferous (pine) woods.

Host: *Schizolachnus pineti* (F.) on *Pinus silvestris*, *Pinus nigra*, *Schizolachnus* sp. on *Pinus mugo*.

Host-specificity: A specialized parasite of *Schizolachnus* -species.

Occurrence: VI, VII, VIII, IX, X.

Economic importance: It was observed to be a rather effective species. Because of its specificity it might be a useful species in control of *Schizolachnus*-aphids.

Note: Mummified aphids are blackish.

Genus *Metaphidius* Starý and Sedlag

Aphidius Nees subg. *Metaphidius* Starý and Sedlag, 1959, Dtsch. ent. Z., N. F. 6: 160 – 161

Type: *Aphidius (Metaphidius) trioxyformis* Starý and Sedlag.

Distribution: Europe (Czechoslovakia, Germany, Austria).

Note: This genus includes parasites of the *Lachnidae*. Pupation inside mummified aphid.

48

1. *M. aterrimus* (Fahringer)

Coelonotus aterrimus Fahringer, 1935, in Schimitschek, Zentralbl. ges. Forstwes. 61: 215

Aphidius (Metaphidius) trioxyformis Starý and Sedlag, 1959, Dtsch. ent. Z., N. F. 161—165

Distribution: Europe (Czechoslovakia, Germany, Austria).

Habitat: Coniferous (pine) forests.

Host: *Cinara* sp. on *Pinus silvestris*.

Host-specificity: A parasite of *Cinara*-species.

Occurrence: VII.

Economic importance: Economically indifferent species.

Note: Mummified aphids are brownish or blackish. The morphological modification of the abdomen of the female enables probably the infestation of lower host instars only.

Genus *Protaphidius* Ashmead

Coelonotus Förster, 1862, Verh. naturh. Ver. preuss. Rheinl. 19: 248 (Preocc.)
   Type-species: *Coelonotus rufus* Förster
Protaphidius Ashmead, 1900, Canad. Ent. 32: 368
   Type-species: *Coelonotus rufus* Förster
Menozzia Goidanich, 1934, Boll. Lab. ent. Bologna 6: 217—227
   Type-species: *Menozzia formicaria* Goidanich

Revision: Starý, 1958, Acta Faun. Ent. Mus. Nat. Pragae 3: 89—93

Distribution: Europe, Far East.

Note: Species of this genus are specialized parasites of *Stomaphis*-species.

1. *P. wissmannii* (Ratzeburg)

Aphidius wissmannii Ratzeburg, 1848, Ichn. d. Forstins. 2: 59
Coelonotus rufus Förster, 1862, Verh. naturh. Ver. preuss. Rheinl. 19: 248
Menozzia formicaria Goidanich, 1934, Boll. Lab. ent. Bologna 6: 217—227

Distribution: Europe.

Habitat: Deciduous woods, parks.

Host: *Stomaphis* sp. on *Quercus* sp.

Host-specificity: Strictly specialized parasite of *Stomaphis*-species.

Occurrence: Probably VII.

Economic importance: Because of its specificity that includes host which may be controlled by chemicals only with difficulties, this species might be a useful parasite in case of serious *Stomaphis*-outbreak.

Note: Mummified aphids are brownish or blackish. Free cocoons, i.e. such mummified aphids from which the aphid skin was removed by ants, are shiny brownish, having appearance of balls.

Because of the morphological adaptation of abdomen that enables the infestation of hosts that live in crevices of rind, this parasite is an example of strictly specialized parasite.

Genus *Diaeretus* Förster

Diaeretus Förster, 1862, Verh. naturh. Ver. preuss. Rheinl. 19: 249
   Type-species: *Aphidius leucopterus* Haliday

   Revision: Starý, 1960, Acta Soc. ent. Čechoslov. 57: 239 – 240

   Distribution: Europe, Far East, ? Nea.

Note: Species of this genus are strictly specialized parasites of Lachnids (*Protolachnus* and allied genera).

1. *D. leucopterus* (Haliday)

Aphidius leucopterus Haliday, 1834, Ent. Mag. 2: 103
Aphidius exspectatus Gautier and Bonnamour, 1936, Bull. Soc. Linn. Lyon, N. S. 5: 74

   Distribution: Europe, Far East.

   Habitat: Coniferous (pine) woods.

   Host: *Protolachnus agilis* (F.) on *Pinus silvestis, P. nigra.*

   Host-specificity: Strictly specialized parasite of *Protolachnus*-species.

   Occurrence: V, VI, VII, VIII, IX, X.

Economic importance: This is an effective parasite of *Protolachnus*-species. In case of host's outbreak it might be an economically valuable species. In conditions of Czechoslovakia it is but an economically indifferent species.

   Note: Mummified aphids are yellowish.

Genus *Aclitus* Förster

Aclitus Förster, 1862, Verh. naturh. Ver. preuss. Rheinl. 19: 248
   Type-species: *Aclitus obscuripennis* Förster

50

Revision: Starý, Beitr. Ent. 9: 184–189

Distribution: Europe (Czechoslovakia, Germany, Hungary), Far East.

1. *A. obscuripennis* Förster

Aclitus obscuripennis Förster, 1862, Ver. naturh. Ver. preuss. Rheinl. 19: 248

Distribution: Europe (Czechoslovakia, Hungary, Germany).

Habitat: It was swept in forest undergrowth.

Host: Unknown in Czechoslovakia. It is known as a parasite of *Anoecia* sp.

Host-specificity: The mode of host life will be probably very important in the host-specificity of this species, similar to *Paralipsis enervis* (Nees).

Occurrence: VI.

Economic importance: It cannot be classified because of too poor material.

Genus *Aphidius* Nees

Incubus Schrank, 1802, Fauna boica 2: 315 (see Opinions).
   Type-species: *Ichneumon aphidum* L.
Aphidius Nees, 1818, Nov. Acta Acad. L. L. C., p. 302
   Type-species: *Bracon picipes* Nees
Theracmion Holmgren, 1872, Ofvers. Svensk. Vet. Akad. Forh., 29(6): 99
   Type-species: *Theracmion arcticus* Holmgren
Aphidius Nees subg. *Euaphidius* Mackauer, 1961, Beitr. Ent. 11: 110
   Type-species: *Aphidius pterocommae* Ashmead

Distribution: Almost cosmop.

Note: Pupation inside mummified aphid. Instar I larva with simple cauda.

1. *A. absinthii* Marshall

Aphidius absinthii Marshall, 1896, in André, Spec. Hym. Eur. d'Alg. 5: 605–606

Distribution: Europe, C. Asia, Far East.

Habitat: This is a typical inhabitant of steppe type habitats. It is rather common in steppe preservations, waste places, ditches, roadsides, everywhere, where *Macrosiphoniella*-species occur.

Host: *Macrosiphoniella absinthii* (L.) on *Artemisia absinthium*, *Macrosiphoniella artemisiae* (B. d. F.) on *Artemisia vulgaris*, *Macrosiphoniella kaufmanni* Börner on *Achillea pontica*, *Achillea millefolium*, *Macrosiphoniella millefolii* (Deg.) on *Achillea millefolium*, *Achillea nobilis*, *Macrosiphoniella pulvera* (Walk.) on *Artemisia maritima*, *Macrosiphoniella stägeri* HRL. on *Centaurea stoebe*, *Macrosiphoniella xeranthemi* Bosh. on *Xeranthemum foetidum*, *Macrosiphoniella* spp. on *Achillea sudetica*, *Matri-*

caria chamomilla, Artemisia campestris, Xeranthemum foetidum, Artemisia vulgaris, Artemisia campestris, Leucanthemum vulgare.

Host-specificity: A typical parasite of *Macrosiphoniella*-species.

Occurrence: VI, VII, VIII.

Economic importance: Economically indifferent species.

Note: Mummified aphids are yellowish. In many cases it was found to be an effective parasite of *Macrosiphoniella*-species.

2. *A. aulacorthi* Starý

Aphidius aulacorthi Starý, 1963, Acta Ent. Mus. Nat. Pragae 35: 601 – 602

Distribution: Europe (Czechoslovakia).

Habitat: Forest (undergrowth) and intermediate habitats.

Host: *Aulacorthum geranii* (Kalt.) on *Erodium cicutarium, Geranium affine, Aulacorthum* spp. on *Vincetoxicum officinale, Naumburgia thyrsiflora, Geranium robertianum, Potentilla argentea, Sanguisorba minor.*

Host-specificity: A parasite of *Aulacorthum*-species.

Occurrence: VI.

Economic importance: Probably economically indifferent species.

Note: Mummified aphids are yellowish.

3. *A. avenae* Haliday

Aphidius (Aphidius) avenae Haliday, 1834, Ent. Mag. 2: 99
Aphidius granarius Marshall, 1896, in André, Spec. Hym. Eur. d'Alg. 5: 579 – 580

Distribution: Europe.

Habitat: It occurs commonly in steppe type habitats, in corn fields namely.

Host: *Sitobium* spp. on *Secale cereale, Hordeum distichon, Avena sativa.*

Host-specificity: A typical parasite of *Sitobium*-species on *Gramineae*, probably also of other aphids on corn *(Rhopalosiphum)*.

Occurrence: VI, VII, VIII.

Economic importance: It is an effective parasite of aphids infesting *Gramineae*, especially corn. Heavy infestations on *Sitobium* attacking *Avena sativa* were observed in submountain districts.

Note: Mummified aphids are yellowish to brown yellowish.

4. *A. caraganae* Starý

Aphidius caraganae Starý, 1963, Acta Ent. Mus. Nat. Pragae 35: 603 – 604

Distribution: Europe (Czechoslovakia, Eur. part of USSR).

Habitat: Forest type habitats (parks).

Host: *Acyrthosiphon caraganae* (Chol.) on *Caragana arborescens.*

Host-specificity: A typical parasite of *Acyrthosiphon caraganae* (Chol.).

5. *A. equiseticola* Starý

Aphidius equiseticola Starý, 1963, Acta Ent. Mus. Nat. Pragae 35: 604–605

Distribution: Europe (Czechoslovakia).

Habitat: Forest undergrowth.

Host: *Sitobium equiseti* Holman on *Equisetum silvaticum.*

Host-specificity: Probably a specialized parasite of the above mentioned host aphid.

Occurrence: VIII, IX, X.

Economic importance: Economically indifferent species.

Note: Mummified aphids are yellow whitish.

6. *A. ervi* Haliday

Aphidius (Aphidius) ervi Haliday, 1834, Ent. Mag. 2: 100
Aphidius (Aphidius) urticae Haliday, 1834, Ent. Mag. 2: 100
Aphidius medicaginis Marshall, 1898, in André, Spec. Hym. Eur. d'Alg. 5bis: 249
Aphidius fumipennis Györfi, 1958, Acta Zool. Hung. 4: 133
Aphidius (Aphidius) smithi Sharma and Subba Rao, 1958, Indian J. Ent. 20: 183–187

Distribution: Pal., Nea., Orient.

Habitat: It occurs in steppe habitats, rarely in edges of parks. Because of its main host occurrence it is rather common in alfalfa and red clover fields.

Host: *Acyrthosiphon caraganae* Chol. on *Caragana arborescens, Acyrthosiphon pisum* (Harris) on *Trifolium montanum, Melilotus albus, Melilotus officinalis, Medicago varia, Medicago sativa, Dorycnium herbaceum, Trifolium pratense, Astragalus* sp., *Vicia faba, Vicia* sp., *Acyrthosiphon spartii* (Koch) on *Sarothamnus scoparius, Acyrthosiphon superbum* Börner on *Seseli hypomarathum, Seseli osseum, Acyrthosiphon* sp. on *Oxytropis pilosa, Astragalus vesicarius, Macrosiphum rosae* (-L.) on *Rosa* sp., *Microlophium evansi* (Theo.) on *Urtica dioica.*

Host-specificity: A typical parasite of *Acyrthosiphon*-species and allied genera *(Microlophium).*

Occurrence: V, VI, VII, VIII, IX, X.

Economic importance: This is a very effective parasite of *Acyrthosiphon pisum* (Harris), a serious pest aphid on forage crops in Czechoslovakia. Almost 90% parasitization was observed during the mentioned aphid outbreak in southern Moravia, in other districts, on *Trifolium,* where there was no aphid outbreak, its effectiveness reached about 60%. In every case this parasite species must be dealt with in case of integrated control of the pea aphid.

53

Note: Mummified aphids are yellowish. The species was studied both in laboratory and field conditions in Czechoslovakia (see Starý, 1962, Zoo. listy 11(3): 265–278).

7. *A. funebris* Mackauer

Aphidius (Aphidius) funebris Mackauer, 1961, Boll. Lab. Ent. Portici 19: 279–280

Distribution: Europe, N. Africa.

Habitat: This is a typical inhabitant of steppe habitats, especially of steppe preservations, waste places, everywhere, where *Dactynotus*-species occur.

Host: *Dactynotus aeneus* HRL. on *Carduus acanthoides*, *Dactynotus campanulae* (Kalt.) on *Campanula* sp., *Campanula rotundifolia*, *Dactynotus cichorii* (Koch) on *Cichorium intybus*, *Centaurea cyanus*, *Crepis biennis*, *Dactynotus cirsii* (L.) on *Cirsium* sp., *Dactynotus jaceae* (L.) on *Centaurea scabiosa*, *Centaurea stoebe*, *Dactynotus muralis* (Bckt.) on *Mycelis muralis*, *Dactynotus obscurus* (Koch) on *Hieracium silvaticum*, *Dactynotus picridis* (F.) on *Picris hieracioides*, *Dactynotus sonchi* (L.) on *Sonchus oleraceus*, *Dactynotus* spp. on *Cirsium* sp., *Carduus* sp., *Hieracium* sp., *Crepis biennis*, *Centaurea* sp., *Carduus crispus*, *Carduus glaucus*, *Campanula sibirica*, *Paczoskia major* Börn. on *Echinops sphaerocephalus*.

Host-specificity: A typical parasite of *Dactynotus*-species and allied genera (*Paczoskia*).

Occurrence: VI, VII, VIII, IX, X.

Economic importance: Economically indifferent species.

Note: Mummified aphids are yellowish to brownish. In many cases this species was observed to be very effective.

8. *A. hieraciorum* Starý

Aphidius hieraciorum Starý, 1962, Bull. ent. Pologne 32: 109–110

Distribution: Europe.

Habitat: Forest undergrowth, gardens.

Host: *Nasonovia nigra* HRL. on *Hieracium silvaticum*, *Nasonovia pilosellae* (Börner) on *Hieracium pilosella*, *Nasonovia ribisnigri* (Mosl.) on *Hieracium* sp., *Nasonovia* spp. on *Hieracium* sp., *Hieracium echioides*, *Hieracium silvaticum*, *Hieracium pilosella*.

Host-specificity: A typical parasite of *Nasonovia*-species.

Occurrence: V, VI, VII.

Economic importance: Economically indifferent species. Nevertheless it might be a useful control of *Nasonovia ribisnigri*. (Mosl.) attacking lettuce in gardens.

Note: Mummified aphids are yellowish to yellow brownish. In many cases it was observed to be a very effective parasite of *Nasonovia*-species on *Hieracium*.

9. *A. hortensis* Marshall

Aphidius hortensis Marshall, 1896, in André, Spec. Hym. Eur. d'Alg. 5: 590—591
Aphidius (Aphidius) berberidis Smith, 1944, Ohio State Univ. Contr. Zoo. Ent.
6: 54—55

Distribution: Europe, Nea.

Habitat: Deciduous woods, parks, orchards.

Host: *Liosomaphis berberidis* (Kalt.) on *Berberis vulgaris*.

Host-specificity: A specialized parasite of *Liosomaphis berberidis* (Kalt.).

Occurrence: VI, VII.

Economic importance: Economically indifferent species.

Note: Mummified aphids are whitish, being placed usually on upper side of leaves.

10. *A. lonicerae* Marshall

Aphidius lonicerae Marshall, 1896, in André Spec. Hym. Eur. d'Alg. 5: 572—573
Aphidius silenes Marshall, 1896, in André, Spec. Hym. Eur. d'Alg. 5: 603—604
Aphidius ulmi Marshall, 1896, in André, Spec. Hym. Eur. d'Alg. 5: 576—577
Aphidius silvaticus Starý, 1962, Bull. ent. Pologne 32: 114—118

Distribution: Europe.

Habitat: A typical species, occurring in forest undergrowth.

Host: *Amphorophora ampullata* Bckt. on *Dryopteris austriaca, Aulacorthum dryopteridis* Holman on *Dryopteris austriaca, Macrosiphum daphnidis* Börner on *Daphne mezereum, Macrosiphum gei* (Koch) on *Geum* sp., *Macrosiphum prenanthidis* Börner on *Prenanthes purpurea, Macrosiphum stellariae* Theo. on *Stellaria holostea*.

Host-specificity: Although all its hosts belong to the *Dactynotinae*, the mode of host life seems to play an important role as all the hosts live in forest undergrowth, having a similar mode of life but having different taxonomic affinity.

Occurrence: V, VI, VII, VIII.

Economic importance: Economically indifferent species.

Note: Mummified aphids are yellowish to yellow whitish.

11. *A. matricariae* Haliday

Aphidius (Aphidius) matricariae Haliday, 1834, Ent. Mag. 2: 103
Aphidius (Aphidius) cirsii Haliday, 1834, Ent. Mag. 2: 101
Aphidius (Aphidius) arundinis Haliday, 1834, Ent. Mag. 2: 106
Aphidius phorodontis Ashmead, 1888, U. S. Natl. Mus. Proc. 11: 662
Aphidius polygoni Marshall, 1896, in André, Spec. Hym. Eur. d'Alg. 5: 602—603
Aphidius lychnidis Marshall, 1896, in André, Spec. Hym. Eur. d'Alg. 5: 607
Aphidius affinis Quilis M. P., 1931, Eos Madrid 7: 48—50
Aphidius arundinis Haliday var. *obscuriforme* Quilis, 1931, Eos Madrid 7: 50—51

Aphidius renominatus Hincks, 1943, Ent. mon. Mag. 79: 44

Aphidius nigriteleus Smith, 1944, Ohio State Univ. Contr. Zoo. Ent. 6: 61–62

Distribution: Pal. Nea.

Habitat: This is a comparatively eurytopic species. Nevertheless, it is believed to occur mainly in steppe and intermediate habitats, penetrating in forest habitats, too.

Host: *Galiobium langei* Börner on *Galium verum, Hydaphias hofmanni* Börner on *Galium verum, Galium molugo, Linosiphon galiophagus* (Wimsh.) on *Galium silvaticum, Linosiphon asperulophagus* Holman on *Asperula odorata, Myzus ajugae* (Schoutt.) on *Ajuga reptans, Ajuga genuensis, Myzus cerasi* (F.) on *Prunus cerasus.*

Host-specificity: This is a parasite of many Aphidine and Myzine aphids.

Occurrence: V, VI, VII.

Economic importance: Economically valuable species, infesting a number of various pest aphids all over its distribution area.

Note: Mummified aphids are yellowish to white yellowish.

12. *A. megourae* Starý

Aphidius megourae Starý, 1965, Acta Faun. Ent. Mus. Nat. Pragae 10: 215–216.

Distribution: Europe (Czechoslovakia, Eur. part of USSR).

Habitat: Meadows, forest meadows.

Host: *Megoura viciae* Bckt. on *Lathyrus pratensis, Vicia cracca.*

Host-specificity: This is probably a specialized parasite of *Megoura viciae* (Bckt.).

Occurrence: VII.

Economic importance: This species was studied carefully in a separate paper (Starý, 1964a). It was ascertained to be very effective, parasitizing about 110 aphids on the average per one female in laboratory. It is a splendid laboratory object, easily bred in a laboratory; because of its high effectiveness it has been used in unnatural host propagation experiments in Czechoslovakia, in which certain pest aphid species like *Acyrthosiphon pisum* (Harris) are used as unnatural hosts. In addition, it was used in biological control of *Megoura viciae* Bckt. experiments on *Vicia faba.*

Note: Mummified aphids are brown yellowish.

13. *A. mirotarsi* Starý

Aphidius mirotarsi Starý, 1963, Acta Ent. Mus. Nat. Pragae 35: 605–606

Distribution: Europe (Czechoslovakia).

Habitat: Steppe habitats.

Host: *Mirotarsus cyparissiae* (Koch) on *Euphorbia cyparissias.*

Host-specificity: This is probably a specialized parasite of the above mentioned host aphid.

Occurrence: VI, VII, VIII.

Economic importance: Economically indifferent species.

Note: Mummified aphids are yellowish.

14. *A. pascuorum* Marshall

Aphidius pascuorum Marshall, 1896, in André, Spec. Hym. Eur. d'Alg. 5: 577–578
Aphidius beltrani Quilis, 1931, Eos Madrid 7: 51–54
Aphidius macropterus Quilis, 1931, Eos Madrid 7: 54–56
Aphidius granarius Marshall var. *pailloti* Quilis, 1931, Eos Madrid 7: 57–58
Aphidius poacearum Starý, 1963, Acta Ent. Mus. Nat. Pragae 35: 608–609

Distribution: Europe.

Habitat: It occurs in steppe habitats, in meadows, corn fields, roadsides, etc.

Host: *Sitobium* sp.

Host-specificity: This is a parasite of some *Gramineae* infesting aphids *(Sitobium, Rhopalosiphum, Metopolophium)*.

Occurrence: VI.

Economic importance: Economically valuable species.

Note: Mummified aphids are yellowish.

15. *A. phalangomyzi* Starý

Aphidius phalangomyzi Starý, 1963, Acta Ent. Mus. Nat. Pragae 35: 607–608

Distribution: Europe (Czechoslovakia).

Habitat: It occurs in steppe type habitats, namely in waste places, long fallow lands, etc., where weeds *(Artemisia)* are common.

Host: *Phalangomyzus oblongus* (Mordv.) on *Artemisia vulgaris*.

Host-specificity: This is a strictly specialized parasite of *Phalangomyzus*-species. It is closely related to the parasites of related aphid genera *(Macrosiphoniella, Dactynotus)* of the *Dactynotinae*.

Occurrence: VI, VII, VIII.

Economic importance: Economically indifferent species.

Note: Mummified aphids are yellowish.

16. *A. picipes* (Nees)

Bracon picipes Nees, 1811, Mag. Ges. naturf. Fr. Berlin 5: 28

Distribution: Europe.

Habitat: It occurs commonly in steppe type habitats, to a lesser degree in intermediate habitats. Because of its main host occurrence, it may be found rather commonly in potato fields.

Host: *Myzodes auctus* (Walk.) on *Cerastium tomentosum*, *Myzodes persicae* Sulz. on *Solanum tuberosum, Papaver dubium, Urtica urens*

57

Host-specificity: A typical parasite of Myzine aphids *(Myzodes, Myzus)*.

Occurrence: V, VI, VII, VIII, IX.

Economic importance: This is an economically valuable species, being a parasite of serious aphid pest.

Note: Mummified aphids are yellow whitish.

17. *A. pterocommae* Ashmead

Aphidius pterocommae Ashmead, 1889, Proc. U. S. Nat. Mus. 11: 659
Aphidius lachni Ashmead, 1889, Proc. U. S. Nat. Mus. 11: 660
Aphidius pterocommae Marshall, 1896, in André, Spec. Hym. Eur. d'Alg. 5: 578 – 579
Aphidius gregarius Marshall, 1872, Ent. mon. Mag. 9: 123

(Note: This is a synonym of *A. cingulatus* Ruthe.)

Distribution: Pal. Nea.

Habitat: Forest and intermediate habitats, irrigating ditches along which *Salix* are grown, parks, avenues along roads, etc.

Host: *Pterocomma pilosum* Bckt. on *Salix caprea*, *Pterocomma salicis* (L.) on *Salix amygdalinus*, *Pterocomma* spp. on *Salix* sp., *Salix caprea*, *Salix cinerea*, *Populus* sp.

Host-specificity: A specialized parasite of *Pterocomma*-species.

Occurrence: IV, V, VI, VII, VIII.,

Economic importance: Economically indifferent species.

Note: Mummified aphids are brownish. The species was found to be an effective parasite of the mentioned aphid species. It probably attacks groups of aphids gradually as mummified aphids were often found in smaller groups.

18. *A. ribis* Haliday

Aphidius (Aphidius) ribis Haliday, 1834, Ent. Mag. 2: 101
Aphidius scabiosae Marshall, 1896, in André, Spec. Hym. Eur. d'Alg. 5: 596 – 597
Aphidius ribis Ashmead, 1898, Wash. Ent. Soc. Proc. 4 (2): 167

Distribution: Europe, Nea.

Habitat: It occurs in forest and intermediate habitats, in parks, orchards, gardens.

Host: *Cryptomyzus ribis* (L.) on *Ribes rubrum*, *Ribes nigra*, *Myzella galeopsidis* (Kalt.) on *Galeopsis tetrahit*.

Host-specificity: A specialized parasite of *Cryptomyzus* and allied genera.

Occurrence: V, VI, VII, VIII.

Economic importance: This is an economically valuable species, being a parasite of serious aphid pest. Nevertheless, its effectiveness is usually high after damage caused by aphid pests on *Ribes*-shrubs.

Note: Mummified aphid are yellow whitish.

19. *A. rosae* Haliday

Aphidius (Aphidius) rosae Haliday, 1833, Ent. Mag. 1: 261
Aphidius rosarum Nees, 1834, Hym. Ichn. aff. Mon. 1: 19
Aphidius aphidivorus Ratzeburg, 1844, Ichn. d. Forstins. 1: 50, 52
Aphidius cancellatus Buckton, 1876, Mon. Brit. Aphid. Vol. 1: 111

Distribution: Europe, ? Nea.

Habitat: Forest and intermediate habitats. It is common in parks, gardens and field shrubs namely, everywhere where *Rosa*-shrubs, usually infested by *Macrosiphum rosae* (L.) occur.

Host: *Macrosiphum rosae* (L.) on *Rosa* spp., *Scabiosa columbaria*, *Dipsacus* sp., *Macrosiphum funestum* (Macch.) on *Rubus* sp.

Host-specificity: A specialized parasite of some *Macrosiphum*-species, *M. rosae* L. namely.

Occurrence: V, VI, VII, VIII.

Economic importance: Economically indifferent species, except maybe in gardens, where *Macrosiphum rosae* (L.) occurs as pest aphid on various cultivated *Rosa*-species and varieties.

Note: The species was found to be comparatively effective, nevertheless, the parasitized aphids were usually found to occur separately. It is probable they spread before being killed by the parasite larva on tops of stems and leaves far from the colony, where they were infested by the parasite female. Mummified aphids are yellowish.

20. *A. rubi* Starý

Aphidius rubi Starý, 1962, Bull. ent. Pologne 32: 112−114

Distribution: Europe (Czechoslovakia, Finland, Eur. part of USSR).

Habitat: It occurs in forest type habitats, in gardens, orchards, forest undergrowth.

Host: *Macrosiphum funestum* (Macch.) on *Rubus* sp.

Host-specificity: This is probably a specialized parasite of *Nectarosiphon rubi* (Kalt.) on *Rubus* and allied aphids on *Rubus* like *Macrosiphum funestum* (Macch.).

Occurrence: VI.

Economic importance: This species may have a certain economic significance in gardens, where cultivated *Rubus*-species are grown.

Note: Mummified aphids are brown yellowish.

21. *A. salicis* Haliday

Aphidius (Aphidius) salicis Haliday, 1834, Ent. Mag. 2: 102
Aphidius dauci Marshall, 1896, in André, Spec. Hym. Eur. d'Alg. 5: 601−602

Distribution: Europe, Far East.

Habitat: It occurs in forest and intermediate habitats, in deciduous woods, parks, gardens.

Host: *Aphidae* ssp. on *Conium maculatum*, *Aphis lambersi* Börn. on *Daucus carota*, *Aphis farinosa* Gmel. on *Salix*, *Cavariella* spp. on *Daucus carota*, *Angelica silvestris*, *Selinum carvifolia*, *Anthriscus silvestris*, *Salix* spp., *Semiaphis dauci* (F.) on *Daucus carota*.

Host-specificity: A typical parasite of *Cavariella*-species and other aphid groups that have a similar mode of life and live on the same plants as the main host aphids.

Occurrence: V, VI, VII.

Economic importance: This species might have a certain significance in gardens as parasite of pest aphids on *Daucus*.

Note: Mummified aphids are yellowish.

22. *A. setiger* Mackauer

Aphidius setiger Mackauer, 1961, Boll. Lab. Ent. Agr. Portici 19: 273–275

Distribution: Europe, Caucasus.

Habitat: Deciduous woods, parks, gardens, avenues.

Host: *Periphyllus villosus* (Htg.) on *Acer platanoides*.

Host-specificity: A specialized parasite of *Periphyllus*-species.

Occurrence: V, VI.

Economic importance: Economically indifferent species.

Note: Mummified aphids are whitish.

23. *A. sicarius* Mackauer

Aphidius sicarius Mackauer, 1961, Boll. Lab. Ent. Agr. Portici 19: 281–282

Distribution: Europe, Siberia.

Habitat: Deciduous woods, clearings, parks.

Hosts: *Calaphis* spp. on *Betula* spp.

Host-specificity: This is a parasite of *Calaphis*-species and some other aphids infesting *Betula*-trees. (*Betulaphis*).

Occurrence. VI, VII, VIII.

Economic importance: Economically indifferent species.

Note: Mummified aphids are white-yellowish.

24. *A. sonchi* Marshall

Aphidius sonchi Marshall, 1896, in André, Spec. Hym. Eur. d'Alg. 5: 585–586

Distribution: Europe.

Habitat: Steppe type habitats, especially waste places, long fallow lands, everywhere where weeds (*Sonchus*) can be found.

Host: *Hyperomyzus lactucae* (L.) on *Sonchus oleraceus, Sonchus asper.*

Host-specificity: A specialized parasite of *Hyperomyzus lactucae* (L.) in steppe habitats.

Occurrence: VI, VII, VIII.

Economic importance: It is a useful parasite because of parasitizing *Hyperomyzus lactucae* (L.), a pest aphid on *Ribes* spp. (primary host plant) and *Lactuca* (secondary host plant).

Note: Mummified aphids are yellowish. It was found to be effective in some cases.

25. *A. tanacetarius* Mackauer

Aphidius tanacetarius Mackauer, 1962, Beitr. Ent. 12: 645 – 646
Aphidius tanaceticola Starý, 1963, Acta Ent. Mus. Nat. Pragae 35: 609 – 610

Distribution: Europe.

Habitat: It occurs in steppe type habitats, in meadows, waste places, long fallow lands, where *Tanacetum* can be found.

Host: *Metopeurum fuscoviride* Stroyan on *Tanacetum vulgare.*

Host-specificity: A strictly specialized parasite of *Metopeurum*-species.

Occurrence: VI, VII.

Economic importance: Economically indifferent species.

Note: Mummified aphids are yellowish. This species was found in some cases to be an effective parasite of the mentioned host.

26. *A. transcaspicus* Telenga

Aphidius transcaspicus Telenga, 1958. Uzbekist. Zool. Zhurn. 2: 55 – 56

Distribution: S. Europe, Asia Minor, Central Asia.

Habitat: This species has been introduced in Czechoslovakia. It was liberated and initially established in plum orchards and in reeds on ponds.

Host: *Hyalopterus pruni* (Geoffr.) on *Prunus domestica, Prunus* spp., *Phragmites communis.*

Host-specificity: This is a specialized parasite of *Hyalopterus pruni* (Geoffr.) and related genera *(Longuiunguis).*

Occurrence: VI (established), VII, VIII, IX, X.

Economic importance: Experiments on its initial establishment have shown that this is a highly effective species, being an important parasite of an aphid pest on plum and peach trees.

Note: Mummified aphids are yellowish or brownish.

Genus *Lysaphidus* Smith C. F.

Aphidius Nees subg. *Lysaphidus* Smith C. F., 1944, Ohio State Univ. Contr. Zoo. Ent. 6: 72
Type-species: *Aphidius (Lysaphidus) adelocarinus* Smith C. F.
Revision: Starý, 1960, Bull. ent. Pologne 30: 357 – 366
Distribution: Europe, Nea.

Note: Pupation inside parasitized aphid.

1. *L. arvensis* Starý

Lysaphidus arvensis Starý, 1960, Bull. ent. Pologne 30: 359 – 361
Distribution: Europe (Czechoslovakia).
Habitat: Steppe habitats (meadows, gardens, pathways, etc.).
Host: *Coloradoa achilleae* HRL. on *Achillea millefolium, Coloradoa tanacetina* (d. Gu.) on *Tanacetum vulgare.*
Host-specificity: A specialized parasite of *Coloradoa*-species.
Occurrence: VI.
Economic importance: Economically indifferent species.
Note: Mummified aphids are yellowish, usually hardly visible on little curled leaves.

2. *L. erysimi* Starý

Lysaphidus erysimi Starý, 1960, Bull. ent. Pologne 30: 361 – 363
Distribution: Europe (Czechoslovakia).
Habitat: Steppe habitats.
Hosts: *Lipaphis erysimi* (Kalt.) on *Erysimum erysimifolium, Pseudobrevicoryne erysimi* Holman on *Erysimum dubium, Erysimum crepidifolium.*
Host-specificity: A parasite of some Myzine aphids – *Lipaphis* and allied genus *Pseudobrevicoryne.*
Occurrence: V, VI, VII, VIII, IX, X.
Economic importance: Economically indifferent species.
Note: Mummified aphids are yellowish. .

Genus *Diaeretellus* Starý

Diaeretellus Starý, 1960, Acta Soc. ent. Čechoslov. 57: 243 – 244.
Type-species: *Aphidius ephippium* Haliday

Distribution: Europe.

Note: Pupation inside parasitized aphid. Species of this genus seem to occur namely in peat-bogs and similar habitats (marshes, etc.).

1. *D. ephippium* (Haliday)

Aphidius ephippium Haliday, 1834, Ent. Mag. 2: 105

Distribution: Europe (Br. Isles, Czechoslovakia, Sweden).

Habitat: Peat-bogs, wet meadows.

Host: Unknown.

Host-specificity: A parasite of *Decorosiphon*-species.

Occurrence: VI, VII, VIII, IX, X.

Economic importance: Economically indifferent species.

2. *D. heinzei* (Mackauer)

Aphidius (Aphidius) heinzei Mackauer, 1959, Dtsch. ent. Z., N. F. 6: 83—84

Distribution: Europe (Germany, Czechoslovakia).

Habitat: Peat bogs, wet meadows, forest undergrowth.

Host: Unknown.

Host-specificity: A parasite of *Decorosiphon*-species.

Occurrence: VI.

Economic importance: Economically indifferent species.

3. *D. macrocarpus* Mackauer

Diaeretellus macrocarpus Mackauer, 1961, Boll. Lab. Ent. Agr. Portici 19: 275—277

Distribution: Europe (Germany, Czechoslovakia).

Habitat: Wet meadows.

Host: Unknown.

Host-specificity: A parasite of *Bacillaphis, Izyphya,* etc.

Occurrence: VI.

Economic importance: Economically indifferent species.

Genus *Diaeretiella* Starý

Diaeretiella Starý, 1960, Acta Soc. ent. Čechoslov. 57: 242—243
Type-species: *Aphidius rapae* M'Intosh

Distribution: Almost cosmop.

Revision: Starý, 1961, Acta Ent. Mus. Nat. Pragae 34: 383 — 397

Note: Pupation inside parasitized aphid.

1. *D. rapae* (M'Intosh)

Aphidius rapae M'Intosh, 1855, Book of the garden, 2, p. 194
Diaeretus chenopodii Förster, 1867, in Kirchner, Catal. Hym. Eur., p. 125
Trioxys piceus Cresson, 1880, U. S. Dept. Agric. Ann. Rpt. for 1879, p. 260
Lipolexis chenopodiaphidis Ashmead, 1889, U. S. Nat. Mus. Proc. 11: 671
Aphidius brassicae Marshall, 1896, in André, Spec. Hym. Eur. d'Alg. 5: 597 — 598
Diaeretus californicus Baker, 1909, Pomona Jl. Ent. 1: 25
Diaeretus nipponensis Viereck, 1911, Proc. U. S. Nat. Mus. 40: 182
Diaeretus obsoletus Kurdjumov, 1913, Rev. Russ. ent. St. Peterburg 13: 25 — 26
Diaeretus napus Quilis M. P., 1931, Eos Madrid 7: 71
Diaeretus croaticus Quilis M. P., 1934, Eos Madrid 10: 8 — 9
Diaeretus plesiorapae Blanchard, 1940, Rev. Chil. Hist. Nat. Santiago 44: 45 — 48
Diaeretus aphidum Mukerji and Chatterjee, 1950, Proc. R. ent. Soc. London (B) 19: 4 — 6

Distribution: Almost cosmopolitan.

Habitat: This is a typical species for steppe habitats. It is rather common in various field habitats namely, in *Brassicaceae*-fields, potato fields and to a lesser degree in sugar-beet fields too. For its occurrence in field habitats it is necessary for its foci to exist in the neighbourhood of cultivated areas. Such foci represent small areas of waste places or long fallow lands, where *Chenopodium* grows, being attacked commonly by *Hayhurstia atriplicis* (L.), which is one of the main hosts of this parasite.

Host: *Brachycaudus helichrysi* (Kalt.) on *Senecio vulgaris*, *Brachycaudus rumexicolens* Patch on *Rumex acetosella*, *Brachycaudus* sp. on *Rumex acetosella*, *Matricaria* sp., *Brevicoryne brassicae* (L.) on *Brassica oleracea* var. *botrytis*, *Sinapis arvensis*, *Alliaria vulgaris*, *Brassica napus*, *Raphanus raphanistrum*, *Dactynotus* sp. on *Crepis biennis*, *Hayhurstia atriplicis* (L.) on *Chenopodium album*, *Atriplex* sp., *Myzodes persicae* Sulz. on *Beta vulgaris*, *Solanum tuberosum*, *Urtica urens*, *Myzodes* sp. on *Cerastium tomentosum*, *Malachium aquaticum*, *Schizaphis scirpi* Kittel on *Typha angustifolia*, *Sitobium* sp. on *Lolium* sp.

Host-specificity: This is comparatively a widely specialized species, whose main hosts however belong clearly in the *Myzinae*, being *Myzodes*, *Hayhurstia* and *Brevicoryne*-species.

Occurrence: V, VI, VII, VIII.

Economic importance: This is a valuable parasite species. Its significance in controlling *Brevicoryne* is important but because of the big quantity and dense colonies of this pest aphid the parasite is unable to reach high effectiveness and to prevent the aphid outbreak (not to speak of predators, etc.). Results other than those of our field observations might be reached by augmentation of the parasite. In field, the parasite is believed to be most important in control of *Myzodes persicae* Sulz. on potato and sugar beet, where its effectiveness is high.

Note: Mummified aphids are yellowish to white yellowish.

Genus *Lysiphlebus* Förster

Lysiphlebus Förster, 1862, Verh. naturh. Ver. Preuss. Rheinl. 19: 248–250
   Type-species: *Aphidius dissolutus* Nees

Revision: 1960 Mackauer, Beitr. Ent. 10: 582–623

Distribution: Pal., Nea., Orient.

Note: Pupation inside mummified aphid.

1. *L. ambiguus* (Haliday)

Aphidius (Aphidius) ambiguus Haliday, 1834, Ent. Mag. 2: 104–105

Distribution: Europe, Caucasus, C. Asia.

Habitat: It occurs namely in forest type and intermediate habitats. It penetrates along rivers and irrigating ditches, where *Salix*-trees grow, also in field habitats.

Host: *Aphis fabae* Scop. on *Eryngium planum*, *Aphis farinosa* Gmel. on *Salix* sp., *Salix repens* var. *rosmarinifolia*, *Salix viminalis*, *Aphis podagrariae* Schrk. on *Aego-podium podagraria*, *Aphis schneideri* (Börner) on *Ribes rubrum*, *Aphis urticata* (F.) on *Urtica dioica*, *Hydaphias* sp. on *Galium molugo*.

Host-specificity: The main hosts of this parasite belong to the genus *Aphis*, although it attacks some other aphids too. As its main host in Czechoslovakia *Aphis farinosa* Gmel. may be mentioned.

Occurrence: V, VI, VII, VIII.

Economic importance: Although this species is valuable as a parasite of a number of pest aphids, it must be classified as economically indifferent because of composition of its hosts in Czechoslovakia.

Note: Mummified aphids are brownish.

2. *L. arvicola* Starý

Lysiphlebus arvicola Starý, 1961, Bull. ent. Pologne 31: 98–100
Lysiphlebus mackaueri Starý, 1961, Acta Faun. Ent. Mus. Nat. Pragae 7: 141
Lysiphlebus (Lysiphlebus) crocinus Mackauer, 1962, Mitt. Dtsch. Ent. Ges. 21: 12–14

Distribution: Europe, C. Asia.

Habitat: This is a typical species occurring in steppe habitats, in steppe preservations and in xerothermic localities of southern and eastern Czechoslovakia.

Host: *Sipha maydis* (Pass.) on *Medicago falcata, Sipha* spp. on *Agropyrum repens, Agropyrum* sp.

Host-specificity: A specialized parasite of *Sipha*-species.

Occurrence: V, VI, VII.

Economic importance: Although its hosts are sometimes pests on corn, they occur mainly on wild grasses. For this reason also their parasite may be classified as an economically indifferent species.

Note: Mummified aphids are yellowish.

3. *L. dissolutus* (Nees)

Bracon dissolutus Nees, 1811, Mag. Ges. naturf. Fr., Berlin 5: 29

Lysiphlebus (Platycyphus) macrocornis Mackauer, 1960, Beitr. Ent. 10: 591 – 592

Distribution: Europe.

Habitat: Forest undergrowth.

Host: Unknown in Czechoslovakia.

Host-specificity: A parasite of root aphids.

Occurrence: V, VI, VII.

Economic importance: Economically indifferent species.

4. *L. fabarum* (Marshall)

Aphidius fabarum Marshall, 1896, in André, Spec. Hym. Eur. d'Alg. 5: 599 – 600

Aphidius cardui Marshall, 1896, in André, Spec. Hym. Eur. d'Alg. 5: 593 – 594

Aphidius gomezi Quilis, 1930, Bol. Pat. Veg. y Ent. Agríc. 1930: 55 – 57

Aphidius janinii Quilis, 1930, Bol. Pat. Vég. y Ent. Agric. 4: 61 – 63

Lysiphlebus fabarum Marshall var. *inermis* Quilis, 1931, Eos Madrid 7: 46

Lysiphlebus innovatus Quilis, 1931, Eos Madrid 7: 39 – 42

Lysiphlebus moroderi Quilis, 1931, Eos Madrid 7: 43 – 45

Distribution: Europe, Asia Minor, N. Africa, Caucasus, C. Asia.

Habitat: This is a common species occurring in steppe habitats of all kinds. In spring it is common as a parasite of various root (root-collar) aphids and then spreads gradually in cultivated areas (sugar beet fields), where it attacks various pest aphids such as the black bean aphid, etc. After emigration of the dioecious aphid species from cultivated areas it may be met again in boundaries of fields, roadsides, etc.

Host: *Aphis craccivora* (Koch) on *Medicago sativa, Onobrychis sativa, Vicia* sp., *Trifolium* sp., *Caragana arborescens, Aphis euphorbiae* (Kalt.) on *Euphorbia cyparissias, Aphis evonymi* (F.) on *Solanum nigrum, Aphis fabae* Scop. on *Urtica urens, Scorzonera parviflora, Matricaria inodora, Carduus rigens, Cirsium* sp., *Matricaria suaveolens, Beta vulgaris, Campanula rapunculoides, Arctium* sp., *Torilis japonica, Chenopodium album, Amaranthus retroflexus, Dahlia variabilis, Carduus* sp., *Cirsium arvense, Rumex crispus, Euonymus europaea, Cirsium palustre, Carduus crispus, Arctium* sp., *Arctium lappa, Chenopodium* sp., *Chenopodium rubrum, Podospermum canum, Lappa major, Callendula officinalis, Anthemis sancti-johannis, Chenopodium bonus-henricus, Carduus acanthoides, Torilis anthriscus, Melandrium album, Aphis intybi* (Koch) on *Cichorium intybus, Aphis lambersi* (Börner) on *Daucus carota, Aphis newtoni* (Theo.) on *Iris variegata, Aphis plantaginis* (Goetze) on *Plantago media, Plantago major, Aphis polygonata* (Nevs.) on *Polygonum convolvulus, Aphis pomi* Deg. on *Malus silvestris, Pirus communis, Aphis poterii* (Börner) on *Sanguisorba minor, Aphis roepkei* (HRL.) on *Potentilla reptans, Aphis rumicis* (L.) on *Rumex* sp., *Aphis salviae* (Walk.) on *Salvia nemorosa, Salvia pratensis, Salvia* sp., *Aphis stachydis* (Mordv.) on *Stachys recta, Aphis taraxacicola* (Börner) on *Taraxacum officinale, Aphis thomasi* (Börner) on *Scabiosa columbaria, Aphis urticata* (F.) on *Urtica urens, Urtica dioica, Aphis vandergooti* (Börner) on *Achillea millefolium, Aphis verbasci* (Schrk.) on *Verbascum* sp., *Verbascum austriacum, Aphis* spp. on *Anthemis maczygia, Rumex* sp., *Rhamnus cathartica, Cirsium arvense, Cirsium* sp., *Impatiens parviflorum, Achillea ptarmica, Arnica sacchaliensis, Eryngium campestre, Galeopsis speciosa, Polygonum convolvulus, Euphorbia* sp., *Epilobium montanum, Trifolium pratense, Caragana arborescens, Torilis japonica, Brachycaudus cardui* (L.) on *Matricaria inodora, Carduus* sp., *Arctium* sp., *Carduus nutans, Arctium lappa, Prunus spinosa, Carduus crispus, Leucanthemum vulgare, Cirsium eriophorum, Brachycaudus rumexicolens* (Patch) on *Rumex acetosella, Brachycaudus tragopogonis* (Kalt.) on *Tragopogon pratense, Tragopogon* sp., *Brachycaudus* spp. on *Prunus domestica, Matricaria* sp., *Carduus acanthoides, Carduus crispus, Prunus persica, Tanacetum vulgare, Dysaphis* sp. on *Arctium lappa, Hyperomyzus lactucae* (L.) on *Sonchus oleraceus, Microsiphum nudum* Holman on *Achillea nobilis, Paczoskia major* (Börner) on *Echinops sphaerocephalus, Pemphigus* sp. on *Helichrysum arenarium, Protaphis carlinae* (Börner) on *Carlina vulgaris, Rhopalosiphum nymphaeae* (L.) on *Ranunculus* sp., *Sitobion avenae* (F.) on *Festuca* sp.

Host-specificity: This is a widely specialized species, its main hosts belonging to the genera *Aphis* and *Brachycaudus*.

Occurrence: V, VI, VII, VIII, IX.

Economic importance: This is a very valuable species, which was observed to reach high effectiveness in many cases of pest aphids infestation. Nevertheless, its significance is sometimes hardly evaluable as it attacks e.g. pest aphid species on economically indifferent host plants. In every case augmentation of this parasite species should be supported.

Note: Mummified aphids are brownish.

The species was successfully bred for a long time under fluorescent light as a parasite of *Aphis fabae* (Scop.) on *Vicia faba* plants in laboratory. Its oviposition is very slow (lasting about 40 seconds) and it attacks gradually all suitable aphid instars (usually instar II is preferred by the ovipositing female) in an aphid group,

5*

reaching high effectiveness in this way. Mummified aphids are usually situated in dense groups, often all the colony being parasitized. It is a deuterrotokous species.

5. *L. fritzmuelleri* Mackauer

Lysiphlebus (Lysiphlebus) fritzmuelleri Mackauer, 1960, Beitr. Ent. 10: 604−605

Distribution: Europe, Siberia.

Habitat: Steppe habitats, especially meadows and boundaries of fields.

Host: *Aphis craccae* (L.) on *Vicia cracca*.

Host-specificity: A strictly specialized species.

Occurrence: VI, VII, VIII.

Economic importance: Economically indifferent species.

Note: Mummified aphids are blackish.

6. *L. hirticornis* Mackauer

Lysiphlebus hirticornis Mackauer, 1960, Beitr. Ent. 10: 606−608

Distribution: Europe.

Habitat: Steppe habitats, especially meadows, boundaries of fields, waste places, fallow lands, everywhere where *Tanacetum vulgare* grows.

Host: *Metopeurum fuscoviride* Stroyan on *Tanacetum vulgare*.

Host-specificity: A strictly specialized species.

Occurrence: VI, VII.

Economic importance: Economically indifferent species.

Note: Mummified aphids are brownish.

7. *L. melandriicola* Starý

Lysiphlebus melandriicola Starý, 1961, Acta Faun. Ent. Mus. Nat. Pragae 7: 135−137

Distribution: Europe (Czechoslovakia).

Habitat: Steppe type habitats, especially meadows.

Host: *Brachycaudus lychnidis* (L.) on *Melandrium album*.

Host-specificity: This is probably a strictly specialized species.

Occurrence: VI, VII, VIII.

Economic importance: Economically indifferent species.

Note: Mummified aphids are brownish.

8. *L. salicaphis* (Fitch)

Trioxys salicaphis Fitch, 1855, N. Y. State Agr. Soc. Trans. 14: 841

Trioxys populaphis Fitch, 1855, N. Y. State Agr. Soc. Trans. 14: 841

Lipolexis salicaphidis Ashmead, 1889, U. S. Nat. Mus. Proc. 11: 671

Aphidius (Diaeretus) laticephalus Telenga, 1953, Trudy Inst. Zool. parazitol. AN Uzb SSR 1: 172−173

Distribution: Pal., Nea.

Habitat: Deciduous woods, avenues, river banks.

Host: *Chaitophorus* sp. on *Populus* sp.

Host-specificity: A typical parasite of *Chaitophorus*-species.

Occurrence: VI.

Economic importance: Economically indifferent species.

Note: Mummified aphids are brownish.

9. *L. thelaxis* Starý

Lysiphlebus thelaxis Starý, 1961, Bull. ent. Pologne 31: 100 – 102

Distribution: Europe.

Habitat: Deciduous woods.

Host: *Thelaxes dryophila* (Schrk.) on *Quercus* sp.

Host-specificity: A specialized parasite of *Thelaxes-species*.

Occurrence: VI, VII.

Economic importance: Economically indifferent species.

Note: Mummified aphids are dark brownish. Its effectiveness is comparatively low.

Genus *Paralipsis* Förster

Paralipsis Förster, 1862, Verh. naturh. Ver. preuss. Rheinl. 19: 248
　　Type-species: *Aphidius enervis* Nees
Myrmecobosca Maneval, 1940, Bull. Soc. Linn. Lyon 9: 9
　　Type: *Myrmecobosca mandibularis* Maneval

Revision: Starý, 1958, Acta Faun. Ent. Mus. Nat. Pragae 3: 85

Distribution: Europe, Far East.

Note: Species of this genus are typical parasites of root aphids. Pupation inside mummified aphid.

1. *P. enervis* (Nees)

Aphidius enervis Nees, 1834, Hym. Ichn. aff. Mon. 1: 26
Myrmecobosca mandibularis Maneval, 1940, Bull. Soc. Linn. Lyon 9: 10 – 13
Myrmecobosca linnei Hincks, 1949, Ent. Tidskr. 1949: 173 – 174

Distribution: Europe.

Habitat: Steppe habitats, boundaries of fields, waste places, long fallow lands.

Host: *Anoecia* sp. on *Agropyrum repens, Aphis roepkei* HRL. on *Potentilla anserina, Brachycaudus ballotae* (Pass.) on *Ballota nigra, Brachycaudus cardui* (L.) on *Carduus crispus*, *Brachycaudus mordwilkoi* HRL. on *Echium* sp., *Brachycaudus* sp. on *Arctium lappa*, *Dysaphis crataegi* (Kalt.) on *Daucus carota*.

Host-specificity: A widely specialized parasite of root aphids.

Occurrence: V, VI, VII, VIII, IX, X.

Economic importance: It is an economically valuable species as it attacks also some pest aphid species *(Dysaphis, Anoecia, Brachycaudus)* that are difficult to control by chemicals on roots of plants (secondary host plants). In many cases it was found to be an effective parasite of many root aphid species.

Note: Mummified aphids are black.

Genus *Monoctonus* Haliday

Aphidius Nees subg. *Monoctonus* Haliday, 1833, Ent. Mag. 1: 261, 487
   Type-species: *Aphidius (Monoctonus) caricis* Haliday
Aphidileo Rondani, 1848, Nuovi ann. Sci. Nat. e Rend. Bologna (29): 14
   Type-species: *Aphidius resolutus* Nees

   Subgenera. *Falciconus* Mackauer, 1959, Senck. biol., Frankfurt M. 40: 180
   Type-species: *Aphidius pseudoplatani* Marshall
Monoctonus s. str., Starý, 1959, Acta Soc. ent. Čechoslov. 56: 241–242
   Type-species: *Monoctonus caricis* (Haliday)
Paramonoctonus Starý, 1959, Acta Soc. ent. Čechoslov. 56: 238–239
   Type-species: *Monoctonus (Paramonoctonus) angustivalvus* Starý
Harkeria Cameron, 1900, Ann. Mag. Nat. Hist. 6: 537
   Type-species: *Harkeria rufa* Cameron

   Revision: Starý, 1959, Acta Soc. ent. Čechoslov. 56: 237–250

   Distribution: Pal., Nea., Orient.

   Note: Pupation inside mummified aphid. Instar I larva with simple caudal appendage.

1. *M. (P.) angustivalvus* Starý

Monoctonus (Paramonoctonus) angustivalvus Starý, 1959, Acta Soc. ent. Čechoslov. 56: 239–241

   Distribution: Europe.

   Habitat: Forest undergrowth.

   Host: *Nasonovia nigra* HRL. on *Hieracium silvaticum*.

   Host-specificity: A parasite of *Nasonovia*-species.

   Occurrence: V, VII, VIII.

   Economic importance: Economically indifferent species.

   Note: Mummified aphids are yellowish.

70

2. M. (M.) caricis (Haliday)

Aphidius (Monoctonus) caricis Haliday, 1833, Ent. Mag. 1: 261, 481

Distribution: Europe (Czechoslovakia, Ireland, Gt. Britain).

Habitat: Meadows in forests, forest undergrowth.

Host: *Sitobium equiseti* Holman on *Equisetum silvaticum.*

Host-specificity: Unclear, probably a more widely specialized parasite.

Occurrence: VI, VII, VII.

Economic importance: Economically indifferent species.

3. M. (M.) cerasi (Marshall)

Aphidius cerasi Marshall, 1896, in André, Spec. Hym. Eur. d'Alg. 5: 607–608

Distribution: Europe.

Habitat: Gardens, deciduous woods.

Host: *Myzodes ligustri* (Mosl.) on *Ligustrum vulgare.*

Host-specificity: A parasite of some leaf-curling species (*Myzodes,* etc.) namely.

Occurrence: VI.

Economic importance: Economically indifferent species. In neighbouring countries it is known as a parasite of certain pest aphid species.

4. M. (M.) crepidis (Haliday)

Aphidius (Monoctonus) crepidis Haliday, 1834, Ent. Mag. 2: 94
Aphidius tuberculatus Wesmael, 1835, Nouv. Mém. Acad. Sci. Bruxelles 9: 80
Monoctonus paludum Marshall, 1896, in André, Spec. Hym. Eur. d'Alg. 5: 548–549

Distribution: Europe.

Habitat: Forest undergrowth, parks, gardens.

Host: *Nasonovia nigra* HRL. on *Hieracium silvaticum, Hieracium* sp., *Hieracium fallax, Nasonovia pilosellae* (Börner) on *Hieracium pilosella, Nasonovia ribisnigri* (Mosl.) on *Hieracium fallax, Hieracium pilosella, Hieracium bauhinii, Hieracium pratense, Lapsana* sp., *Nasonovia* spp. on *Hieracium echoides, Hieracium junceum.*

Host-specificity: A parasite of *Nasonovia*-species.

Occurrence: V, VI.

Economic importance: This might be an economically useful species, parasitizing *Nasonovia ribisnigri* (Mosl.) in gardens (known from Gt. Britain).

Note: Mummified aphids are yellow brownish.

5. M. (M.) nervosus (Haliday)

Aphidius (Monoctonus) nervosus Haliday, 1833, Ent. Mag. 1: 488
Monoctonus birói Györfi, 1958, Acta Zool. Hung. 4: 132
Monoctonus breviantennalis Starý, 1959, Acta Soc. ent. Čechoslov. 56: 242–243

Distribution: Europe (Czechoslovakia, Great Britain).

Habitat: Forest undergrowth.

Host: *Impatientinum balsamines* (Kalt.) on *Impatiens nolli-tangere.*

Host-specificity: Probably a strictly specialized species.

Occurrence: VI, VII, VIII.

Economic importance: Economically indifferent species.

Note: Mummified aphids are reddish brown.

6. *M. (F.) pseudoplatani* (Marshall)

Aphidius pseudoplatani Marshall, 1896, in André, Spec. Hym. Eur. d'Alg. 5: 582 – 583

Distribution: Europe.

Habitat: Forest type habitats – deciduous woods, parks, avenues.

Host: *Drepanosiphum platanoides* (Schrk.) on *Acer pseudoplatanus.*

Host-specificity: A strictly specialized parasite.

Occurrence: V, VI, VII.

Economic importance: Economically indifferent species.

Note: Mummified aphids are yellowish.

Genus *Trioxys* Haliday

Aphidius Nees subg. *Trioxys* Haliday, 1833, Ent. Mag. 1: 261, 488

Type-species: *Aphidius cirsii* Curtis

Neuropenes Provancher, 1886, Add. Faun. Canad. Hym., pp. 151 and 153

Type-species: *Neuropenes ovalis* Provancher

Subgenera: *Betuloxys* Mackauer, 1960, Beitr. Ent. 10: 139

Type-species: *Trioxys compressicornis* Ruthe

Binodoxys Mackauer, 1960, Beitr. Ent. 10: 141

Type-species: *Aphidius (Trioxys) angelicae* Haliday

Pectoxys Mackauer, 1960, Beitr. Ent. 10: 154 – 155

Type-species: *Trioxys (Trioxys) macroceratus* Mackauer

Trioxys s. str., Smith, 1944. Ohio State Univ. Contr. Zoo. Ent. 6: 85

Type-species: *Aphidius cirsii* Curtis

Revision: Mackauer, 1959, Beitr. Ent. 9: 144 – 179, 1960, Beitr. Ent. 10: 137 – 160

Distribution: Almost cosmop.

Note: Pupation inside mummified aphid. Instar I larva with simple cauda.

1. *T. (B.) acalephae* (Marshall)

Aphidius acalephae Marshall, 1896, in André, Spec. Hym. Eur. d'Alg. 5: 608 – 609

Trioxys (Trioxys) urticae Mackauer, 1959, Beitr. Ent. 9: 171 – 173

Distribution: Pal.

Habitat: This is a comparatively eurytopic species, but judging from the number of samples collected the steppe habitat seems to be preferred.

Hosts: *Aphis craccae* (L.) on *Vicia* sp., *Aphis craccivora* (Koch) on *Robinia pseudoacacia*, *Caragana arborescens*, *Trifolium* sp., *Onobrychis sativa*, *Aphis cystisorum* Htg. on *Laburnum vulgare*, *Aphis euphorbiae* (Kalt.) on *Euphorbia cyparissias*, *Aphis fabae* Scop. on *Arctium* sp., *Cirsium arvense*, *Aphis farinosa* Gmel. on *Salix* sp., *Aphis mordwilkiana* (Dobrowlj.) on *Rubus* sp., *Aphis salviae* (Walk.) on *Salvia pratensis*, *Salvia nemorosa*, *Aphis spiraephaga* Müller on *Spiraea arguta*, *Aphis urticata* (F.) on *Urtica dioica*, *Aphis* spp. on *Epilobium montanum*, *Rumex conglomeratus*.

Host-specificity: A typical parasite of *Aphis*-species.

Occurrence: V, VI, VII, VIII.

Economic importance: Economically valuable species.

Note: Mummified aphids are brownish.

2. T. (B.) angelicae (Haliday)

Aphidius (Trioxys) angelicae Haliday, 1833, Ent. Mag. 1: 489
Trioxys placidus Gautier, 1922, Bull. Soc. ent. France 1921: 302
Trioxys granatensis Quilis M. P., 1931, Eos Madrid 7: 74−76
Trioxys boscai Quilis M. P., 1931, Eos Madrid 7: 83−84
Trioxys fumariae Quilis, 1931, Eos Madrid 7: 81−83
Trioxys obscuriformis Quilis, 1931, Eos Madrid 7: 77−78
Trioxys wollastonii Cabrera, Mackauer, 1963, Eos Madrid 38: 439−440
Trioxys angelicae mediterraneus Mackauer, 1960, Beitr. Ent. 10: 142−143

Distribution: Europe, Asia Minor.

Habitat: It is a typical species occurring in forest type and intermediate habitats, from where it spreads to steppe habitats in the neighbourhood.

Host: *Acyrthosiphon caraganae* Chol. on *Caragana arborescens*, *Aphis cognatella* Jones on *Evonymus europaea*, *Aphis craccivora* Koch on *Caragana arborescens*, *Aphis fabae* Scop. on *Campanula rapunculoides*, *Arctium* sp., *Scorzonera parviflora*, *Evonymus europaea*, *Philadelphus coronarius*, *Chenopodium* sp., *Beta vulgaris*, *Aphis farinosa* Gmel. on *Salix* sp., *Myzodes persicae* Sulz. on *Papaver dubium*, *Aphis pomi* Deg. on *Crataegus monogyna*, *Malus silvestris*, *Aphis salviae* (Walk.) on *Salvia verticillata*, *Aphis sambuci* (L.) on *Sambucus nigra*, *Aphis spiraephaga* Müller on *Spiraea* sp., *Spiraea arguta*, *Aphis viburni* Scop. on *Viburnum opulus*, *Aphis* sp. on *Tropaeolum majus*, *Malachium aquaticum*, *Epilobium* sp., *Impatiens nolli-tangere*, *Brachycaudus helichrysi* (Kalt.) on *Persica vulgaris*, *Brachycaudus* sp. on *Rumex acetosella*, *Ceruraphis eriophori* (Walk.) on *Viburnum lantana*, *Dysaphis devecta* (Walk.) on *Malus silvestris*, *Dysaphis* spp. on *Crataegus oxyacantha*, *Rhopalosiphum padi* (L.) on *Padus racemosa*.

Host-specificity: The main hosts of this species belong to the genus *Aphis* but it attacks a number of other aphid groups too *(Ceruraphis, Dysaphis, Myzodes, Brachycaudus)*.

Occurrence: V, VI, VII.

Economic importance: This is an economically valuable species attacking a number of aphid pests.

Note: This species was bred under fluorescent light on *Aphis fabae* Scop. infesting *Vicia faba* plants in laboratory. Length of oviposition 5—8 seconds. Lower instar aphids (II) are clearly preferred by the ovipositing female. The oviposition is gradual being practically interrupted only by short resting periods. The attacked aphids respond in producing a small drop of wax material from siphuncles, the sucking however being uninterrupted. The accessory apparatus enables apparently to attack only lower instar aphids as the attack on adult aphids was usually unsuccessful according to our observations.

Mummified aphids are brownish, being found usually in dense groups. This is caused by the host instar preference by the ovipositing female that attacks a group of an aphid progeny which include a certain number of suitable instars.

3. *T. (T.) auctus* (Haliday)

Aphidius (Trioxys) auctus Haliday, 1833, Ent. Mag. 1: 489

Distribution: Europe.

Habitat: Meadows.

Host: Unknown.

Host-specificity: As far as it is known this species attacks *Rhopalosiphum oxyacanthae* (Schrk.) in Italy.

Occurrence: VI.

Economic importance unknown.

4. *T. (T.) betulae* (Marshall)

Trioxys betulae Marshall, 1896, in André, Spec. Hym. Eur. d'Alg. 5: 553—554

Distribution: Europe.

Habitat: Deciduous woods, clearings.

Host: *Symydobius oblongus* (v. Heyd.) on *Betula* sp.

Host-specificity: A specialized parasite of *Symydobius oblongus* (v. Heyd.).

Occurrence: VIII.

Economic importance: Economically indifferent species.

Note: Mummified aphids are brownish.

5. *T. (B.) brevicornis* (Haliday)

Aphidius (Trioxys) brevicornis Haliday, 1833, Ent. Mag. 1: 491
Aphidius (Trioxys) minutus Haliday, 1833, Ent. Mag. 1: 491

Distribution: Europe.

Habitat: Steppe and intermediate habitats.

Host: *Cavariella* spp. on *Daucus carota, Hyadaphis* sp. on *Conium maculatum, Hyadaphis bupleuri* Börner on *Bupleurum falcatum, Staegeriella necopinata* (Börner) on *Galium verum.*

Host-specificity: A parasite of certain Myzine genera *(Cavariella, Hyadaphis, Staegeriella).*

Occurrence: VI, VII.

Economic importance: Economically indifferent species, except parasitization of *Cavariella*-species in gardens.

Note: Mummified aphids are yellowish brown.

6. *T. (B.) centaureae* (Haliday)

Aphidius (Trioxys) centaureae Haliday, 1833, Ent. Mag. 1: 490

Distribution: Europe.

Habitat: A typical species of steppe habitats, to a lesser degree it penetrates into deciduous woods.

Host: *Dactynotus aeneus* HRL. on *Carduus crispus, Carduus* sp., *Dactynotus campanulae* (Kalt.) on *Campanula* sp., *Dactynotus cichorii* (Koch) on *Crepis biennis, Cichorium intybus, Dactynotus jaceae* (L.) on *Centaurea scabiosa, Centaurea stoebe, Dactynotus muralis* (Bckt.) on *Mycelis muralis, Dactynotus obscurus* (Koch) on *Hieracium* sp., *Macrosiphoniella artemisiae* (B. d. F.) on *Artemisia vulgaris, Macrosiphoniella millefolii* (Deg.) on *Achillea millefolium, Macrosiphoniella tanacetaria* (Kalt.) on *Chrysanthemum leucanthemum, Microlophium evansi* (Theo.) on *Urtica dioica.*

Host-specificity: It is a typical parasite of *Dactynotus* and *Macrosiphoniella*-species, it attacks to a lesser degree also other aphid groups *(Microlophium).*

Occurrence: VI, VII, VIII.

Economic importance: Economically indifferent species.

Note: Mummified aphids are brownish.

7. *T. (T.) cirsii* (Curtis)

Aphidius cirsii Curtis, 1831, Brit. Entom. 8: 383
Aphidius (Trioxys) aceris Haliday, 1833, Ent. Mag. 1: 490
Aphidius resolutus Nees, 1834, Hym. Ichn. aff. Mon. 1: 24–25

Distribution: Europe.

Habitat: Deciduous woods, parks, avenues.

Host: *Drepanosiphon platanoidis* (Schrk.) on *Acer pseudoplatanus.*

Host-specificity: A strictly specialized parasite.

Occurrence: VI, VII, VIII.

Economic importance: Economically indifferent species.

Note: Mummified aphids are yellowish.

8. *T. (T.) falcatus* Mackauer

Trioxys (Trioxys) falcatus Mackauer, 1959, Beitr. Ent. 9: 164 – 165

Distribution: Europe, Caucasus.

Habitat: Deciduous woods, parks, avenues, shrubs in fields.

Host: *Periphyllus villosus* (Htg.) on *Acer platanoides, Acer campestre.*

Host-specificity: A specialized parasite of *Periphyllus*-species.

Occurrence: IV, V, VI.

Economic importance: Economically indifferent species.

Note: Mummified aphids are brownish.

9. *T. (B.) genistae* Mackauer

Trioxys (Binodoxys) genistae Mackauer, 1960, Beitr. Ent. 10: 143 – 146

Distribution: Europe (Czechoslovakia, Germany).

Habitat: Steppe type habitats, waste places, meadows.

Host: *Aphis genistae* Scop. on *Genista* sp.

Host-specificity: A specialized parasite of the mentioned host aphid.

Occurrence: VI.

Economic importance: Economically indifferent species.

Note: Mummified aphids are brownish.

10. *T. (T.) glaber* Starý

Trioxys (Trioxys) glaber Starý, 1966, Acta Ent. Mus. Nat. Pragae (in press)

Distribution: Europe (Czechoslovakia).

Habitat: Steppe habitats.

Host: *Aphis galii-scabri* Schrk. on *Asperula cynanchica.*

Host-specificity: A specialized parasite of the mentioned host aphid.

Occurrence: VI.

Economic importance: Economically indifferent species.

Note: Mummified aphids are brownish.

11. *T. (B.) hortorum* Starý

Trioxys (Betuloxys) hortorum Starý, 1960, Acta Soc. ent. Čechoslov. 57: 365 – 367

Distribution: Europe (Czechoslovakia).

Habitat: Deciduous woods, parks.

Host: *Tinocallis platani* (Kalt.) on *Ulmus effusa.*

Host-specificity: This is probably a very specialized species.

76

Occurrence: V.

Economic importance: Economically indifferent species.

Note: Mummified aphids are yellowish.

12. *T. (T.) humuli* **Mackauer**

Trioxys (Trioxys) humuli Mackauer, 1960, Beitr. Ent. 10: 152—154

Distribution: Europe (Czechoslovakia, Germany).

Habitat: Intermediate habitats (hop gardens).

Host: *Phorodon humuli* (Schrk.) on *Humulus lupulus*.

Host-specificity: A specialized parasite.

Occurrence: VI.

Economic importance: Little has been known on the effectiveness of this species, which is a parasite of serious pest aphid.

13. *T. (P.) macroceratus* **Mackauer**

Trioxys (Pectoxys) macroceratus Mackauer, 1960, Beitr. Ent. 10: 155—156

Distribution: Europe (Czechoslovakia, Gt. Britain).

Habitat: Meadows.

Host: *Aphis podagrariae* Schrk. on *Aegopodium podagraria*.

Host-specificity: This is probably a specialized parasite.

Occurrence: VI.

Economic importance: Economically indifferent species,

14. *T. (T.) pallidus* **(Haliday)**

Aphidius (Trioxys) pallidus Haliday, 1833, Ent. Mag. 1: 489
Aphidius callipteri Marshall, 1896, in André, Spec. Hym. Eur. d'Alg. 5: 609—610
Trioxys pulcher Gautier and Bonnamour, 1924, Bull. Soc. ent. France 1924: 43—44

Distribution: Europe, C. Asia.

Habitat: Forest type habitats — deciduous woods, parks, orchards.

Host: *Chromaphis juglandicola* (Kalt.) on *Juglans regia, Eucallipterus tiliae* (L.) on *Tilia* sp., *Myzocallis carpini* (Koch) on *Carpinus betulus, Tuberculoides annulatus* (Htg.) on *Quercus* sp.

Host-specificity: A typical parasite of a number of *Callaphidinae* occurring in forest type habitats.

Occurrence: V, VI, VII, VIII.

Economic importance: Economically indifferent species. Nevertheless, it was found to be useful in other countries where some of the above mentioned aphids occur as serious pests.

Note: Mummified aphids are yellowish, occurring usually as single specimens on leaves (lower side).

15. *T. (T.) pannonicus* Starý

Trioxys (Trioxys) pannonicus Starý, 1960, Bull. Soc. ent. Mulhouse 1960: 93−96

Distribution: Europe (Czechoslovakia, Hungary).

Habitat: Steppe type habitats, steppe districts of eastern and southern Czechoslovakia.

Host: *Titanosiphon artemisiae* (Koch) on *Artemisia campestris*.

Host-specificity: A strictly specialized species.

Occurrence: VI, VII.

Economic importance: Economically indifferent species.

Note: Mummified aphids are brownish.

16. *T. (T.) parauctus* Starý

Trioxys (Trioxys) parauctus Starý, 1960, Acta Soc. ent. Čechoslov. 57: 367−368

Distribution: Europe (Czechoslovakia).

Habitat: Steppe habitats, meadows, waste places, long fallow lands.

Hosts: *Hydaphias* sp. on *Galium verum, Galium* sp.

Host-specificity: A strictly specialized parasite of *Hydaphias* spp.

Occurrence: V, VI, VII.

Economic importance: Economically indifferent species.

Note: Mummified aphids are light brownish.

17. *T. (T.) phyllaphidis* Mackauer

Trioxys (Trioxys) phyllaphidis Mackauer, 1961, Boll. Lab. Ent. Agr. Portici 19: 286−288

Distribution: Europe (Czechoslovakia).

Habitat: Deciduous woods and forests *(Fagus)*, namely of the Carpathian mountains district.

Host: *Phyllaphis fagi* (L.) on *Fagus silvatica*.

Host-specificity: A strictly monophagous species.

Occurrence: VII, VIII.

Economic importance: This is a valuable species, a parasite of a serious pest aphid, heavy outbreaks of which were often observed in Czechoslovakia. Although its effectiveness observed was comparatively low in comparison with masses of aphids during an outbreak, this is the only parasite known to attack this aphid. Only alate specimens of aphids were observed to be mummified, which fact is of importance for the spread of the parasite.

Note: Mummified aphids are yellowish.

18. *T. (T.) spinosus* Starý

Trioxys (Trioxys) spinosus Starý, 1963, Boll. Lab. Ent. Agr. Portici 21: 212−214

Distribution: Europe (Czechoslovakia).

Habitat: Steppe habitats.

Host: *Semiaphis dauci seselii* Börner on *Seseli osseum*.

Host-specificity: A strictly specialized species.

Occurrence: VI.

Economic importance: Economically indifferent species.

Note: Mummified aphids are yellowish.

Genus *Lipolexis* Förster

Lipolexis Förster, 1862, Verh. naturh. Ver. preuss. Rheinl. 19: 249
 Type-species: *Lipolexis gracilis* Förster
Gynocryptus Quilis, 1931, Eos Madrid 7: 27−28
 Type: *Gynocryptus pieltaini* Quilis

Distribution: Europe, Far East, Phillippines.

Note: Pupation inside mummified aphid. Instar I larva has 2 long lateral oblique appendages besides the cauda.

1. *L. gracilis* Förster

Lipolexis gracilis Förster, 1862, Verh. naturh. Ver. preuss. Rheinl. 19: 249
Gynocryptus pieltaini Quilis, 1931, Eos Madrid 7: 27−28

Distribution: Europe, Far East.

Habitat: Steppe and intermediate habitats, it penetrates also in edges of woods and in orchards.

Host: *Anoecia* sp. on *Cornus sanguinea, Aphis bupleuri* (Börner) on *Bupleurum falcatum, Aphis craccae* (L.) on *Vicia cracca, Aphis craccivora* (Koch) on *Medicago sativa, Onobrychis sativa, Aphis euphorbiae* (Kalt.) on *Euphorbia cyparissias, Aphis fabae* Scop. on *Beta vulgaris, Cirsium* sp., *Centaurea cyanus, Cirsium arvense, Aphis intybi* (Koch) on *Cichorium intybus, Aphis newtoni* Theo. on *Iris variegata, Aphis origani* (Pass.) on *Origanum vulgare, Aphis plantaginis* (Goetze) on *Plantago* sp., *Aphis polygonata* (News.) on *Polygonum aviculare, Aphis salviae* (Walk.) on *Salvia pratensis, Salvia nemorosa, Salvia* sp., *Aphis taraxacicola* (Börner) on *Taraxacum officinale, Aphis* sp. on *Rumex* sp., *Peucedanum alsaticum, Cirsium* sp., *Brachycaudus* sp. on *Tanacetum, Carduus* sp., *Arctium lappa, Prunus domestica, Cirsium vulgare, Chrysanthemum maritimum, Brachycaudus cardui* (L.) on *Carduus* sp., *Brachycaudus helichrysi* (Kalt.) on *Melandrium album, Prunus persica, Brachycaudus mordwilkoi* HRL. on *Echium vulgare, Myzus cerasi* (F.) on *Prunus avium, Prunus cerasus*

79

Host-specificity: The main hosts of this species belong to the genera *Aphis* and *Brachycaudus,* to a lesser degree to the *Myzinae (Myzus)*.

Occurrence: V, VI, VII, VIII.

Economic importance: This is a valuable species, parasitizing a number of pest aphids. Because of its wide specialization it is able to occur in spring as a parasite of various root aphids (root collar) in boundaries of fields, roadsides, etc. from where it spreads gradually to cultivated areas, where pest aphids immigrate during the season.

Note: Mummified aphids are brownish.

Key to genera, subgenera and species (♀♀)

This is an original key to genera, subgenera and species of all determinable species ascertained in Czechoslovakia. When determining the species it is necessary to deal with their ecological characteristics, or to use more detailed papers (references) in doubtful cases. It is necessary to have in mind that the species that might be found in Czechoslovakia, as they were found in the neighbouring countries, are not included in the key. In many cases *(Praon, Aphidius)* it is necessary to have a series of reared specimens at hand if the species is to be identified exactly, otherwise the variability of different species can hardly be examined; in these cases, too, the host represents a valuable additional character.

The key includes all European genera and subgenera to allow at least the generic identification of all species found. The abbreviation Fig. concerrs the key — figures on plates II – IX.

1	Apterous. . . . . see *Diaretellus* Starý (108)
—	Wings fully developed 2
2 (1)	Median vein developed throughout, separating radial cell 1 from median cell (Fig. 21) 3
—	Median vein effaced frontally or entirely, radial cell 1 and median cell 1 confluent; venation often reduced behind basal vein (Figs. 19, 20, 24) 22
3 (2)	Interradial veins effaced (Fig. 31) 4
—	Both interradial veins developed (Fig. 27) 13
4 (3)	Propodeum smooth. Ovipositor sheaths sparsely haired (Fig. 45). Pupation under parasitized aphid in a separate cocoon. Distr.: Pal., Nea. . . . . genus: *Praon* Haliday 5
—	Propodeum more or less distinctly areolated (Fig. 84), ovipositor sheaths densely haired (Fig. 35). Pupation inside parasitized aphid. Distr.: Europe . . . genus: *Areopraon* Mackauer [*A. lepelleyi* (Waterston)].

80

5 (4) Scape, pedicel and F_1 entirely yellow 6

— Scape, pedicel and F_1 brownish, sometimes apex of pedicel and base of F_1 yellowish 9

6 (5) Antenna 20—22 segmented. Lateral lobes of mesoscutum densely pubescent. (Parasites of *Dactynotus* spp.) . . . . . . *P. dorsale* (Haliday)

— Antennae with a smaller number of segments. Lateral lobes of mesoscutum densely pubescent or with hairless areas 7

7 (6) Antennae 18—19 segmented. Lateral lobes of mesoscutum densely pubescent or with hairless areas 8

— Antennae 16—18 segmented. Lateral lobes of mesoscutum with hairless areas. (Parasites of *Therioaphis* spp.) . . . . . *P. exoletum* (Nees)

8 (7) Lateral lobes of mesoscutum densely pubescent. (Parasites of *Protolachnus* spp.) . . . . *P. bicolor* Mackauer

— Lateral lobes of mesoscutum with big oval hairless areas. (Parasites of *Eucallipterus, Tuberculoides* spp.) . . . . . . *P. flavinode* (Haliday)

9 (5) Antennae 14—17 segmented 10

— Antennae 17—20 segmented 12

10 (9) Lateral lobes of mesoscutum with big oval hairless areas. Antennae 14—17 segmented. (Parasites of *Aphis* ssp.) . . *P. abjectum* (Haliday)

— Lateral lobes of mesoscutum densely pubescent 11

11 (10) Thorax brown black, abdomen (except tergite 1 and spot on tergite 2) dark brown. Hairs on middle flagellar segments nearly equal to width of the segment. (Parasites of *Nasonovia*) . . . . . . *P. pubescens* Starý

— Lower part of thorax, tergite 1 and apex of abdomen (except ovipositor sheaths) yellow. Hairs on middle flagellar segments shorter than width of the segment. (Parasites of *Macrosiphum rosae*) . . *P. rosaecola* Starý

12 (9) Lateral lobes of mesoscutum densely pubescent. (Rather polyphagous species) . . . . *P. volucre* (Haliday)

— Lateral lobes of mesoscutum with oval hairless areas (Parasites of *Macrosiphoniella* sp.) . . . . *P. absinthii* Bignell

13 (3) Ovipositor sheaths and ovipositor straight or slightly curved upwards (Fig. 55). Antennae 11-segmented. Abdomen lanceolate. Pupation inside the mummified aphid. Distr.: Pal., Nea., Aeth. . . . .
. . . . genus: *Ephedrus* Haliday 14

— Ovipositor sheaths curved downwards, rather broadened, deltoid and trifid at the extremity (Fig. 62). Ovipositor curved downwards. Antennae 18-segmented. Abdomen rounded. Pupation inside the parasitized aphid. Distr.: Europe . . . . . genus: *Toxares* Haliday (*T. deltiger Hal.*)

14 (13) Propodeum coarsely and irregularly rugose, carinae prominent and sometimes not entirely distinct (Fig. 68). Ovipositor sheaths very wide, strongly narrowing to the apex; entire surface densely pubescent (Fig. 103) . . . (subg. *Lysephedrus* Starý) . . *E. (L.) validus* (Haliday)

–	Propodeum smooth, areolated, neighbourhood of carinae sometimes slightly sculptured (Fig. 46). Ovipositor sheaths comparatively long and narrow, with sparse hairs (Fig. 55). F_1 longer or equal to F_2. Flagellar segments with various numbers of rhinaria, sparsely haired. 15
15 (14)	Radial abscissa 2 shorter than interradial vein (Fig. 22) . . . . *E. (E.) persicae* Froggatt
–	Radial abscissa 2 equal or distinctly longer than interradial vein 1 (Figs. 21, 27) 16
16 (15)	Radial abscissa 2 equal to interradial vein 1 (Fig. 27) 17
–	Radial abscissa 2 distinctly longer than interradial vein 1 (Fig. 21). . 18
17 (16)	F_1 long and slender, $^1/_3$ longer than F_2. F_1 with 0, $F_2 - 1$, $F_3 - 2$ rhinaria (Fig. 106). F_1 brown yellow to yellow . . . . *E. (E.) lacertosus* (Haliday)
–	F_1 stout, about $^1/_6$ longer than F_2. F_1 with 5, $F_2 - 7$ rhinaria. F_1 black, yellowish at base . . . . *E. (E.) campestris* Starý
18 (17)	Tergite 1 scarcely longer than wide at spiracles (Fig. 75), gradually dilated to the apex. Radial cell 2 very long . . . . *E. (E.) brevis* Stelfox
–	Tergite 1 distinctly longer than wide at spiracles, with deep lateral impressions before the apex (Fig. 92) 19
19 (18)	F_1 $^1/_3$ longer than F_2. F_1 with 0, $F_2 - 2$, $F_3 - 2$ rhinaria. F_1 and part of F_2 yellowish . . . . *E. (E.) cerasicola* Starý
–	F_1 only slightly longer or equal to F_2. Flagellar segments with different numbers of rhinaria. F_1 yellowish at base only, the rest of flagellum dark. 20
20 (19)	F_1 with $5 - 7$, $F_2 - 7$, $F_3 - 7$ rhinaria. Praeapical and apical segments distinctly separated (Fig. 98) . . . . *E. (E.) plagiator* (Nees)
	F_1 with 1, $F_2 - 2$, $F_3 - 3$ rhinaria. Praeapical and apical segments distinctly separated to form a club (Fig. 102) 21
21 (20)	Praeapical and apical segments forming a club. Tergite 1 with richly developed carinae . . . . *E. (E.) minor* Stelfox
–	Praeapical and apical segments distinctly separated from each other. Tergite 1 with strongly developed central and lateral longitudinal carinae . . . . *E. (E.) nacheri* Quilis M. P.
22 (2)	Radial and median cells confluent, distinctly completed by 2nd interradial vein along their external margin (2nd interradial vein sometimes nearly colourless but distinct) (Figs. 24, 15) 23
–	Radial and median cells confluent, open, not completed by interradial vein 2 along their external margin (Fig. 20). 80
23 (22)	Pterostigmal cell distinctly complete. Eyes small, antennae moniliform. Notaulices distinct at base as slight rugosities. Propodeum smooth. Abdomen rounded. Tergite 1 transverse. Distr.: Europe . . . . genus: *Aclitus* Förster (*A. obscuripennis* Förster)

–	Pterostigmal cell distinctly incomplete 24
24 (23)	Confluent radial and median cells distinctly separated on lower margin by intermedian + median veins (Fig. 24) 25
–	Confluent radial and median cells on the lower margin open – the rest of median vein visible only under the 2nd interradial vein (Fig. 15) . . 70
25 (24)	Abdominal segments beginning with the 4th remarkably tubiform and telescopic. Pupation under mummified aphid. Distr.: Europe, Far East . . . . genus: *Protaphidius* Ashmead (*P. wissmannii* Ratz.)
–	Abdominal segments of normal shape, abdomen lanceolate or rounded 26
26 (25)	Ovipositor sheaths curved downwards, ploughshare-shaped (Fig. 71), or slender, gradually narrowing to the apex (Fig. 79). Note: Wing venation variable. Pupation inside mummified aphid. Distr.: Pal., Nea . . . . genus: *Monoctonus* Haliday 27
–	Ovipositor sheaths slightly curved upwards (Fig. 69) 32
27 (26)	Propodeum distinctly areolated, with central areola (Fig. 66). Ovipositor sheaths stout, ploughshare-shaped, or slender, gradually narrowing to the apex. 28
–	Propodeum with 2 divergent carinae in lower part (Fig. 100). Ovipositor sheaths slender, gradually narrowing to the apex (Fig. 79) . . . . (subg. *Paramonoctonus* Starý) . . . *M. (P.) angustivalvus* Starý
28 (27)	Ovipositor sheaths stout, ploughshare-shaped (Fig. 71) (subg. *Monoctonus* s. str.) 29
–	Ovipositor sheaths slender, gradually narrowing to the apex (subg. *Falciconus* Mackauer) . . . . *M. (F.) pseudoplatani* (Marshall)
29 (28)	Antennae 15 – 16-segmented . . . . . . *M. (M.) nervosus* (Haliday)
–	Antennae 13 – 14-segmented 30
30 (29)	Propodeum with large and very wide central areola (Fig. 66) . . . . *M. (M.) crepidis* (Haliday)
–	Propodeum with narrow central areola (Fig. 56) 31
31 (30)	Tergite 1 yellow to yellow brown, slightly sculptured or almost smooth along the distinct central carina and its bifurcation . . . . *M. (M.) cerasi* (Marshall)
–	Tergite 1 dark brown, coarsely rugose, sometimes with central carina and its bifurcation distinct . . . . *M. (M.) caricis* (Haliday)
32 (26)	Carinae on propodeum forming large, wide pentagonal areola (sometimes poorly visible in the longitudinal part) (Figs. 81, 89) 58
–	Carinae on propodeum forming very narrow, small, central areola (Fig. 37). Pupation inside mummified aphid. Distr.: Almost cosmopolitan . . . . genus: *Aphidius* Nees 33
33 (32)	Tentorio-ocular line equal to half of intertentorial line or longer . . 34

–	Tentorio-ocular line equal to $^1/_3$ of intertentorial line or somewhat longer 44
34 (33)	Antennae with 14 – 15-segments. (Parasites of *Periphyllus* spp.). . . . . *A. setiger* (Mackauer)
–	Antennae with another number of segments 35
35 (34)	Temple $^1/_3 - ^1/_2$ narrower than transverse eye-diameter 36
–	Temple almost equal to transverse eye-diameter (usually as $^1/_5 - ^1/_6$) 37
36 (35)	Antennae 18 – 20-segmented. (Parasites of *Pterocomma* -spp.). . . . . *A. pterocommae* Ashmead
–	Antennae 16 – 17-segmented. (Parasites of *Metopeurum*-spp.). . . . . *A. tanacetarius* Mackauer
37 (34)	Antennae 15 – 16-segmented. (Parasites of *Hyalopterus*-spp.) . . . . *A. transcaspicus* Telenga
–	Antennae with another number of segments 38
38 (37)	Antennae 16 – 17-segmented. (Parasites of *Macrosiphoniella*-spp.) . . . . *A. absinthii* Marshall
–	Antennae with another number of segments 39
39 (38)	Antennae 17 – 18-segmented. (Parasites of *Macrosiphum*-spp., *M. rosae* Hal. namely) . . . . *A. rosae* Haliday
–	Antennae with another number of segments 40
40 (39)	Antennae 18 – 19-segmented 41
–	Antennae with 19 – 21-segments 42
41 (40) .	Face yellow. Metacarpus equal to pterostigma. Radial abscissa 1 equal to 2. Coloration yellowish, with brown upper part of head and thorax (Parasites of *Amphorophora, Macrosiphum*-spp.) . . . . *A. lonicerae* Marshall
–	Face mostly brownish or with yellowish spots, seldom yellowish. Metacarpus usually shorter than width of pterostigma. Radial abscissa 1 longer than 2. Coloration mostly brownish, with yellowish clypeus, prothorax and tergite 1. (Parasites of *Dactynotus*-spp.) . . . . *A. funebris* Mackauer
42 (40)	Antennae 19 – 20-segmented 43
–	Antennae 20 – 21-segmented. (Parasites of *Phalangomyzus*-spp.). . . . . *A. phalangomyzi* Starý
43 (42)	Wings normal. Radial abscissa 1 more or less equal to 2. Apex of abdomen yellowish, ovipositor sheaths brown. (Parasites of *Megoura*-spp.) . . . . *A. megourae* Starý
–	Wings ample. Radial abscissa 1 usually $^1/_3$ longer than 2. Apex of abdomen dark brown. (Parasites of *Nectarosiphum, Macrosiphum*-spp.) . . . . *A. rubi* Starý
44 (33)	Antennae 18 – 19-segmented. (Parasites of *Acyrthosiphon, Microlophium*, etc.) . . . . *A. ervi* Haliday

–	Antennae with 12 – 18-segments	45
45 (44)	Antennae 17 – 18-segmented	46
–	Antennae with 12 – 17-segments	50

46 (45) Tergite 1 with sharply prominent central carina. (Parasites of *Sitobium, Rhopalosiphum*-spp.). . . . . *A. avenae* Haliday

– Tergite 1 with slightly prominent central carina 47

47 (46) Metacarpus distinctly shorter than pterostigma. (Parasites of *Acyrthosiphon caraganae* Chol.) . . . . *A. caraganae* Starý

– Metacarpus subequal to the length of pterostigma 48

48 (47) Scape, pedicel and F_1 dark brown. Coloration dark brown. (Parasites of *Myzodes*-spp.) . . . . *A. picipes* (Nees)

– Scape, pedicel and F_1 mostly yellowish. Body with yellowish coloration **49**

49 (46) Face entirely brownish. Thorax brown, prothorax sometimes lighter. (Parasites of some *Sitobium*-spp.) . . . . *A. equiseticola* Starý

– Face mostly yellowish. Thorax mostly yellowish anteriorly. (Parasites of *Mirotarsus*-spp.) . . . . *A. mirotarsi* Starý

50 (45)	Antennae 16 – 17-segmented	51
–	Antennae with 12 – 16-segments	52

51 (50) Scape, pedicel and F_1 mostly brownish with yellowish patterns. Antennae reaching half of abdomen. Coloration black brown, clypeus yellowish, prothorax and tergite 1 lighter. Legs brown yellowish. (Parasites of *Aulacorthum*-spp.). . . . . *A. aulacorthi* Starý

– Scape, pedicel and F_1 mostly yellowish, with brownish patterns. Antennae subequal to body length. Coloration brownish, face and clypeus, prothorax and tergite 1 yellowish. Legs yellow. (Parasites of *Sitobium, Metopolophium*-spp.) . . . . *A. pascuorum* Marshall

52 (50)	Antennae 15 – 16-segmented	53
–	Antennae with 12 – 15-segments	55

53 (52) Abdomen brown, tergite 1 and base of tergite 2 yellowish. (Parasites of *Nasonovia*-spp.) . . . . *A. hieraciorum* Starý

– Abdomen yellowish to orange, more or less darkened in the central part. Ovipositor sheaths dark 54

54 (53) Face and usually all the lower part of head, including mouthparts, yellow. (Parasites of *Cryptomyzus*-spp.) . . . . *A. ribis* Haliday

– Face brownish, sometimes with yellowish patterns. Clypeus, lower portion of genae and mouthparts yellowish. (Parasites of *Hyperomyzus*-spp.) . . . . *A. sonchi* Marshall

55 (52)	Antennae 14 – 15-segmented	56
–	Antennae with 12 – 14-segments	57

56 (55) Apex of abdomen brown. (Parasites of a number of myzine and aphidine aphids) . . . . *A. matricariae* Haliday

–	Apex of abdomen yellow or yellowish, ovipositor sheaths brown. (Parasites of *Liosomaphis berberidis* Kalt.) . . . . . . *A. hortensis* Marshall
57 (55)	Antennae 13–14-segmented, slender. Intermedial + median veins often colourless and hardly visible. (Parasites of *Betulaphis*) . . . . *A. sicarius* Mackauer
–	Antennae 12–13-segmented, thickened to the apex. Interradial + median veins coloured. (Parasites of *Cavariella, Semiaphis*) . . *A. salicis* Haliday
58 (32)	Tergite 7 with small tubiform prong at base (Fig. 80). Pupation inside parasitized aphid. Distr.: Europe. . . . . genus: *Metaphidius* Starý + Sedlag [*M. aterrimus* (Fahringer)]
–	Tergite 7 without small tubiform prong at base. Pupation inside parasitized aphid. Distr.: Pal., Nea., Orient. . genus: *Pauesia* Quilis M.P. . 59
59 (58)	Antennae 28–30-segmented . . . . *P. grossa* (Fahringer)
–	Antennae with smaller number of segments 60
60 (59)	Flagellar segments (except for 1 and 2 that are sometimes yellowish or brownish on the lower side) entirely black 61
–	Flagellar segments (except for 1 and 2) black; apical part of flagellum except for apical segment whitish . . . . . . *P. infulata* (Haliday)
61 (60)	Ovipositor sheaths wide, stout, only slightly narrowed to the apex and slightly curved upwards (Fig. 33) 64
–	Ovipositor sheaths of another shape (Figs. 36, 50, 57) 62
62 (61)	Antennae 16–17-segmented. Mesoscutum falling comparatively vertically to prothorax without covering it when viewed laterally (Fig. 107). Central areola on propodeum very large, rounded. Ovipositor sheaths very slender, only slightly curved upwards, strongly narrowing to the apex (Fig. 57) . . . . *P. unilachni* (Gahan)
–	Antennae with greater number of segments. Mesoscutum strongly raised above prothorax and covering it when viewed laterally (Fig. 65). Central areola on propodeum of another shape. Ovipositor sheaths of another shape (Figs. 36, 50) 63
63 (62)	Tergite 1 in the basal half – before spiracular tubercles – strongly curved (Fig. 93). Ovipositor sheaths slender, stout, strongly curved upwards, parallel-sided, bluntly pointed at the apex (Fig. 50) . . . . *P. picta* (Haliday)
–	Tergite 1 in the basal half – before the spiracular tubercles – only slightly curved (Fig. 95). Ovipositor sheath slender, weaker, less curved downwards, narrowed at the base and apex (Fig. 36) . . . . *P. laricis* (Haliday)
64 (61)	Rami of the central carina on propodeum more or less developed, more or less completing the large central areola that differs strongly by its declivity from the neighbourhood (Fig. 89). Gena narrower than longitudinal eye-diameter 65

–   Rami of the central carina on propodeum very little developed, sometimes hardly differing from the slightly rugose neighbourhood, completing large central areola that does not practically differ by its declivity from the neighbourhood. Gena as wide as half of longitudinal eye-diameter (Fig. 76) . . . . *P. maculolachni* (Starý)

65 (64)   Tergite 1 with lateral impressions beyond spiracular tubercles and strongly dilating to the apex; twice (or nearly so) as long as it is wide at spiracles; comparatively flat beyond spiracular tubercles, with central longitudinal impression and with one tuberculiform protuberance on either side (Fig. 97) 66

–   Tergite 1 beyond prominent spiracular tubercles without or only with very small lateral impressions; legs dilating to the apex and a little wider at the apex than width at spiracles; comparatively flat beyond spiracular tubercles (Fig. 105) 68

66 (65)   Antennae 18–20-segmented . . . . *P. silvestris* (Starý)

–   Antennae 21–22-segmented 67

67 (66)   Pterostigma brown, yellowish at base. Propodeum with very large and concave central areola . . . . *P. pini* (Haliday)

–   Pterostigma brown. Propodeum with less concave and smaller central areola . . . . *P. abietis* (Marshall)

68 (65)   Thorax black, prothorax brownish to brown yellow. Tergite 1 and 2 brown, the rest of abdomen black. Antennae 20–21-segmented . . . . *P. juniperorum* (Starý)

–   Thorax with more distributed yellow, rufous or brownish coloration. Abdomen yellowish and brown. Antennae 18–19 and 19–20-segmented 69

69 (68)   Prothorax yellow-rufous. Mesoscutum yellow-rufous, very rarely a little obscured on the lobes. The rest of thorax yellow, rufous or obscured to black. Antennae 18–19–segmented. Tergite 1 somewhat dilating beyond spiracular tubercles towards apex . . . . *P. cupressobii* (Starý)

–   Prothorax yellow. Mesoscutum black, on the base in the neighbourhood of notaulices and sometimes throughout the effaced notaulices on the disc yellowish to light brown. Mesopleurae beneath tegulae slightly yellowish, the rest of thorax black. Antennae 19–20-segmented. Tergite 1 almost parallel-sided . . . . *P. piceaecollis* (Starý)

70 (24)   Tergite 1 with more or less developed central tubercles only, without central carina or coarse rugosities (Fig. 74). Tentorio-ocular line almost or absolutely equal to intertentorial line. Propodeum smooth or with 2 short divergent carinae in lower part. Anterior prong of valvulae 2 normal (Fig. 53). Pupation inside mummified aphid. Distr.: Almost cosmopolitan . . . . genus: *Lysiphlebus* Förster 71

| | Tergite 1 with more or less distinct central carina, more or less rugose (Fig. 83). Tentorio-ocular line distinctly shorter than intertentorial line, usually as $^1/_3$. Anterior prong of valvulae 2 large and appearing flat from side (Fig. 43). Pupation inside mummified aphid. Distr.: Europe, Nea., Neotr. . . . . genus: *Lysaphidus* Smith C. F. . 79 |

– Tergite 1 with more or less distinct central carina, more or less rugose (Fig. 83). Tentorio-ocular line distinctly shorter than intertentorial line, usually as $^1/_3$. Anterior prong of valvulae 2 large and appearing flat from side (Fig. 43). Pupation inside mummified aphid. Distr.: Europe, Nea., Neotr. . . . . genus: *Lysaphidus* Smith C. F. . 79

71 (70) The rest of interradial vein and rest of median vein distinct (Fig. 15) . 72

– Only radial vein distinct, the rest of interradial and median veins effaced (Fig. 16) 77

72 (71) Middle and apical segments of flagellum almost square. Propodeum with two distinct divergent carinae (Fig. 100) in lower part. Antennae 15 – 16-segmented . . . . *L. dissolutus* (Nees)

– Middle and apical segments of flagelum distinctly longitudinal, usually twice as long as wide. Propodeum smooth. Antennae 12 – 13, 13 – 14-segmented 73

73 (72) Metacarp distinctly shorter than pterostigma, not reaching wing apex. (Parasites of *Aphis craccae*) . . . . *L. fritzmuelleri* Mackauer

– Metacarp distinctly longer than pterostigma, reaching wing apex . . 74

74 (73) Tibiae with semierected hairs (Fig. 85). (Parasites of *Metopeurum* spp.) . . . . *L. hirticornis* Mackauer

Tibiae with adpressed hairs 75

75 (74) Lower and apical margin of wing with long hairs (Fig. 17) . . . . *L. ambiguus* (Haliday)

– Lower and apical margin of wing without hairs (Fig. 15) 76

76 (75) Antennae 12 – 13-segmented. (Widely polyphagous species) . . . . *L. fabarum* (Marshall)

– Antennae 13 – 14-segmented. (Parasites of *Brachycaudus lychnidis*) . . . . *L. melandriicola* Starý

77 (71) Tibiae with adpressed hairs. (Parasites of *Chaitophorus* spp.) . . . . *L. salicaphis* (Fitch)

– Tibiae with semierected hairs 78

78 (77) Apex of abdomen yellowish. Ovipositor sheaths dark brown. Tergite 1 with keeliform tubercles before spiracular tubercles (Fig. 86). (Parasites of *Sipha* spp.) . . . . *L. arvicola* Starý

– Apex of abdomen and ovipositor sheaths black brown. Tergite 1 with widely triangular tuberculiform protuberance before spiracles (Fig. 73). (Parasites of *Thelaxes* spp.) . . . . *L. thelaxis* Starý

79 (70) Temple as wide as transverse eye-diameter. Eyes large (Fig. 47). Antennae 13-segmented, only slightly thickened towards apex, somewhat longer than head, thorax and tergite 1 together. Flagellar segment 2 somewhat more than three times longer than width at apex. Head black brown; face, clypeus, mouthparts and genae yellow. Antennae brown, basal half

88

yellowish. Thorax brown black, prothorax mostly yellow . . . .
. . . . *L. arvensis* Starý

— Temple about $^1/_3$ wider than transverse eye-diameter. Eyes small (Fig. 49). Antennae 12-segmented, somewhat thickened towards apex, as long as head and thorax combined. Head black brown; face, clypeus and mouthparts yellowish. Antennae brown; scape, pedicel and part of flagellar segment 1 yellowish. Thorax black brown, prothorax sometimes lighter
. . . . *L. erysimi* Starý

80 (22) Radial vein pointlike (Fig. 32). Pterostigma large, triangular, strongly sclerotized. Legs stout. Pupation inside mummified aphid. Distr.: Europe, Far East . . . . genus: *Paralipsis* Förster [*P. enervis* (Nees)]

— Radial vein distinctly developed, always longer, never pointlike. Legs normal 81

81 (80) Ovipositor sheaths curved downwards, terminal abdominal sternite sometimes with 2 prongs (Figs. 91, 109) 82

— Ovipositor sheaths straight or slightly curved upwards, terminal abdominal sternite without posterior prongs (Figs. 63, 69, 45) 103

82 (81) Terminal abdominal sternite with two prongs . . . .
genus: *Trioxys* Haliday 83

— Terminal abdominal sternite without prongs. 101

83 (82) Tergite 1 with primary (= spiracular) and secondary tubercles, the latter being sometimes poorly visible because nearly fused with primary tubercles (Fig. 99). . . . . (subg. *Binodoxys* Mackauer) . 84

— Tergite 1 with primary tubercles only (Fig. 96) 89

84 (83) Prongs with 7−8 stout long hairs on the dorsal side (Fig. 44) . . . . .
. . . . *T. (B.) centaureae* (Haliday)

— Prongs with a smaller number (usually 5) of hairs on the dorsal side (Fig. 91). 85

85 (84) Prongs almost straight, only slightly curved at apex (Fig. 91) 86

— Prongs strongly curved at apex (Fig. 58) 88

86 (85) Distance between primary and secondary tubercles equal or longer than width at spiracles (Fig. 94) . . . . *T. (B.) angelicae* (Haliday)

— Distance between primary and secondary tubercles shorter than width at spiracles 87

87 (86) Primary tubercles poorly developed, tergite 1 gradually dilating to the apex. . . . . *T. (B.) genistae* Mackauer

— Primary tubercles distinct, the part of tergite 1 between primary and secondary tubercles almost parallel-sided . *T. (B.) acalephae* (Marshall)

88 (85) Face brown. Primary and secondary tubercles very closely related . . .
. . . . *T. (B.) brevicornis* (Haliday)

— Face yellow. Primary and secondary tubercles in a longer distance (Fig. 101) . . . . *T. (B.) heraclei* (Haliday)

89

89 (83)	Prongs curved beyond the centre, with some spines dilated at base (Fig. 60). Tergite 1 almost parallel-sided, primary tubercles hardly visible (subgenus: *Betuloxys* Mackauer) . . . . . . *T. (B.) hortorum* Star̃ý
–	Prongs slightly curved to nearly straight (Figs. 77, 42) 90
90 (89)	Primary tubercles situated at the first third (Fig. 39). Prongs of variable length. Ovipositor sheaths normal, apical hairs normal or dilated at base (Subgenus: *Trioxys* s. str.) 91
–	Primary tubercles situated near the half of the tergite (Fig. 104). Prongs remarkably long, without apical hairs. Ovipositor sheaths rather long, with stout brush-like bristles at the inner side (Fig. 41) (subg. *Pectoxys* Mackauer) . . . . *T. (P.) macroceratus* Mackauer
91 (90)	Apex of prongs with 2 simple bristles (Fig. 64) 92
–	Apex of prongs with 1 – 2 dilated or 1 claw-shaped bristles 95
92 (91)	Antennae 12-segmented. Tergite 1 with lateral longitudinal carinae, between them striated to about the level of spiracular tubercles (Fig. 96) . . . . *T. (T.) auctus* (Haliday)
–	Antennae 11-segmented. Tergite 1 smooth 93
93 (92)	Ovipositor sheaths and prongs yellow brownish, easily distinguishable from the central dark part of abdomen. Prongs with 3 – 4 hairs on dorsal side. (Fig. 111) . . . . *T. (T.) glaber* Starý
–	Other combinations of characters 94
94 (93)	Ovipositor sheaths and prongs dark brown as the central part of abdomen. Prongs with 5 hairs on dorsal side (Fig. 42) . . . . *T. (T.) parauctus* Starý
–	Ovipositor sheaths brown, prongs yellow, central part of abdomen brown. Prongs with 6 strong hairs on dorsal side (Fig. 110) . . . . *T. (T.) spinosus* Starý
95 (91)	Apex of prongs with 1 – 2 dilated bristles (Figs. 38, 77) 96
–	Apex of prongs with 1 claw-shaped bristle 100
96 (95)	Apex of prongs with 2 dilated bristles (Figs. 38, 51) 97
–	Apex of prongs with 1 dilated bristle (Fig. 77) 99
97 (96)	Prongs with 6 – 7 long hairs (Fig. 51) 98
–	Prongs almost hairless (Fig. 38) . . . . . *T. (T.) falcatus* Mackauer
98 (97)	Pterostigma widely triangular, metacarp shorter than width of pterostigma (Fig. 26). . . . . *T. (T.) pannonicus* Starý
–	Pterostigma triangular, metacarp distinctly longer than width of pterostigma . . . . *T. (T.) cirsii* (Curtis)
99 (96)	Hairs (3) on dorsal side of prongs oblique. Ovipositor sheaths with dense hairs on inner and apical part . . . *T. (T.) phyllaphidis* Mackauer
–	Hairs (3) on dorsal side of prongs perpendicular. Ovipositor sheaths with sparse hairs on inner and apical parts (Fig. 77) *T. (T.) humuli* Mackauer

100 (95) Prongs relatively wide, gradually narrowing to the apex, uniformly, arcuated (Fig. 112). (Parasites of *Symydobius* spp.) *T. (T.) betulae* Marshall

– Prongs narrower, almost straight – only slightly arcuated at apex (Fig. 72). (Parasites of *Eucallipterus* and *Tuberculoides*). . . . .
. . . . *T. (T.) pallidus* (Haliday)

101 (82) Radial vein longer than $^2/_3$ of its possible length so that pterostigmal cell nearly complete (Fig. 23). Ovipositor sheaths slightly curved downwards, their upper part more strongly sclerotized (Fig. 109). Pupation inside mummified aphid. Distr.: Europe, Far East . . . .
. . . . genus: *Lipolexis* Förster (*L. gracilis* Föster)

– Radial cell never longer than $^2/_3$ of its possible length; pterostigmal cell distinctly incomplete (Fig. 16). Ovipositor sheaths slightly curved downwards, more or less ploughshare-shaped (Fig. 71) or clawed (Fig. 82) or slender (Fig. 79) 102

102 (101) Tergite 1 always longer than wide at spiracles. Ovipositor sheaths triangular, ploughshare-shaped (Fig. 71) or slender, gradually narrowing to the apex (Fig. 79) . . . . see: *Monoctonus* Haliday

– Tergite 1 square (Fig. 90). Ovipositor sheaths triangular, clawed (Fig. 82). Pupation inside mummified aphid. Distr.: Europe, C. Asia . . . . . .
. . . . genus: *Monoctonia* Starý

103 (81) Notaulices entirely effaced. Propodeum with more or less distinct wide central areola (Fig. 67). Pupation inside mummified aphid. Distr.: Europe, Far East . . . . . . genus: *Diaeretus* Förster [*D. leucopterus* (Haliday)]

– Notaulices at least at the base distinct. 104

104 (103) Propodeum distinctly areolated, with small central areola 105

– Propodeum smooth or with 2 divergent carinae in the lower part . . .
. . . . see: *Lysiphlebus* Förster

105 (104) Head nearly square (Fig. 61). Notaulices deep and distinct throughout. Pupation inside parasitized aphid in a separate cocoon. Distr.: Europe
. . . . genus: *Dyscritulus* Hincks [*D. planiceps* (Marshall)]

– Head transverse. Notaulices more or less deep but distinct at the ascendent part only 106

106 (105) Intermedian vein (fused with part of median vein entirely effaced) (Fig. 18). Pupation inside mummified aphid. Distr.: cosmopolitan . . .
. . . . genus: *Diaeretiella* Starý [*D. rapae* (M'Int.)]

– Intermedian vein (fused with part of median vein) distinct, somewhat less coloured than the radial vein (Fig. 19). Pupation inside mummified aphid . . . . genus: *Diaeretellus* Starý . . 107

107 (106) Apterous (Males winged). . . . . *D. ephippium* (Haliday)

– Wings fully developed 108

108 (107)   Antennae 14−(15)-segmented. Middle flagellar segments at least twice as long as wide . . . . *D. macrocarpus* Mackauer

—   Antennae 13−14-segmented. Middle flagellar segments at most 1.5 times as long as wide . . . . *D. heinzei* (Mackauer)

Genera not included in the key: *Calaphidius* Mackauer, *Tanytrichophorus* Mackauer

V.

Distribution

Geographic distribution of aphidiids is determined by a number of various factors.

The development of the group is the reason that today several centres of development may be recognized, from which apparently the parasites spread to other areas. The development of parasites is connected in many ways also with the development of hosts, with their requirements on the environment, dispersion, etc.

The development of the host group, of aphids, is connected with conditions of the moderate climate in the northern hemisphere. In this climate zone also the greatest number of aphids is distributed. In other areas, besides the cosmopolitan species, certain aphid groups developed in a tropical climate zone (oriental region).

In the zoogeographical classification of aphid, it is necessary to deal with the two circumstances (after Holman, in litt.).*)

1. It is typical for aphids as plant parasites that in the development of species and higher taxonomic units, the ecological isolation, conditioned by gradual adaptation on certain group or plant species, or a part of plant, has played a rather important role. For this reason the distribution of aphids is mainly determined by the distribution of suitable plant species.

2. The mode of aphid spread and their ability to spread over big distances due to air streams, in connection with their large quantity too, is the reason that aphids are gradually able to spread also on small groups of their host plants far beyond the boundaries of their continuous area of distribution.

In this connection the main direction of air streams is important. Only oceans and high mountains seem to represent a serious bar.

Usually the distribution area of an aphid species does not cover the distribution area of the host plant, the temperature conditions on northern and southern boundaries, humidity of the environment namely in forest species, etc. being the main factors.

---

*) The distribution of aphids in Czechoslovaka is classified after MS Pintera and Holman, Fauna of ČSSR (*Aphidoidea*).

As the importance of the host × parasite relationship and ecological factors is not of the same degree, requirements of the host and parasites on the environment are various in their distribution area, the problem is rather complicated by the above mentioned and other factors.

The unsatisfactory world research of the group does not also enable us to know the factors influencing the distribution of many species, as in a number of territories data are completely lacking.

Although the above mentioned difficulties exist in the classifying of the *Aphidiidae,* it is believed that in the classification of the distribution of the *Aphidiidae* it is advantageous to use the results obtained in the study of host specificity influencing factors (see chapter VII).

In connection with the existence of aphidiids in various zoogeographical zones, the following faunistic complexes may be recognized in Europe, connected with the individual types of zones: 1. Tundra, 2. Coniferous forest, 3. Deciduous forest, 4. Steppe and wood steppe, 5. Semi-desert.

Distribution of aphidiids in Czechoslovakia.

In Czechoslovakia the deciduous forest zone and the steppe zone (Pannonian district) are contiguous. Besides, in a number of mountain districts (coniferous and mixed woods) many representatives of boreoalpine fauna may be met.

A more detailed division of habitats of different zones is dealt with in chapter VII.

For the northern and mountain coniferous forest zone, aphidiid species of the genus *Pauesia* are typical. They are parasites of Lachnid species that live on *Picea, Larix* and *Abies.* Some of them seem to be typical inhabitants of these habitats, such as the parasite of *Todolachnus abieticola,* while a number of parasite species penetrate from these habitats to the lowlands.

Furthermore, a number of species that are typical for forest undergrowth may be met here too. They are parasites of a number of aphids: *Aphidius lonicerae* Marsh., etc.

Typical, closely related to the above mentioned type of habitat, are peat-bogs. For them the species of the genus *Diaeretellus* are characteristic, occurring sometimes also in wet forest meadows, etc.

The *Pinus* and *Juniperus* forest or shrubs are rather untypical as to the zoogeographical zones as they occur in a number of habitats in the whole Palearctic region. There are certain parasites, which are connected with *Pinus*-aphids [*Praon bicolor* Mack., *Diaeretus leucopterus* (Hal.), *Pauesia unilachni* (Gahan), etc.], while usually the *Cinara* species occurring on *Pinus,* are attacked by *Pauesia*-species which attack also other *Cinara* species on other conifers.

The undergrowth of *Pinus*-woods depends closely on the environmental habitats (mountain, deciduous woods, steppe, etc.) so that it does not represent typical habitats for parasites.

A big part of the aphidiid fauna is connected with a deciduous forest. They are

94

either parasites of aphids on trees and shrubs. [*Aphidius hortensis* Marshall, *Aphidius pterocommae* Ashmead, *Aphidius rosae* Haliday, *Aphidius setiger* Mackauer, *Aphidius sicarius* Mackauer, *Ephedrus cerasicola* Starý, *Ephedrus persicae* Froggatt, *Ephedrus plagiator* (Nees), *Lysiphlebus ambiguus* (Haliday), *Lysiphlebus salicaphis* (Fitch), *Lysiphlebus thelaxis* Starý, *Monoctonus pseudoplatani* (Marshall), *Praon abjectum* (Haliday), *Praon flavinode* (Haliday), *Protaphidius wissmannii* (Ratzeburg), *Toxares deltiger* (Haliday), *Trioxys angelicae* (Haliday), *Trioxys cirsii* (Haliday), *Trioxys falcatus* Mackauer, *Trioxys pallidus* (Haliday), *Trioxys phyllaphidis* Mackauer], or aphids in undergrowth [*Aphidius aulacorthi* Starý, *Aphidius hieraciorum* Starý, *Ephedrus lacertosus* (Haliday), *Monoctonus angustivalvus* Starý, *Monoctonus crepidis* (Haliday), *Praon pubescens* Starý, *Monoctonus nervosus* (Haliday), etc.].

Many species are also able to spread far over the boundaries of this zone — to the north or in the steppe and wood steppe zone.

The steppe fauna of parasites is rather numerous. It includes parasites of aphids that live primarily in the true steppe districts, but today have spread also over a great part of cultivated steppe zone.

As typical members of the steppe parasite fauna many parasite species may be mentioned: *Aphidius absinthii* Marshall, *Aphidius ervi* Haliday, *Aphidius funebris* Mackauer, *Aphidius phalangomyzi* Starý, *Aphidius sonchi* Marshall, *Aphidius tanacetarius* Mackauer, *Ephedrus campestris* Starý, *Ephedrus nacheri* Quilis M. P., *Diaeretiella rapae* (M'Int.), *Lipolexis gracilis* Förster, *Lysiphlebus arvicola* Starý, *Lysiphlebus fabarum* (Marshall), *Lysiphlebus fritzmuelleri* Mackauer, *Lysiphlebus hirticornis* Mackauer, *Lysaphidus erysimi* Starý, *Lysaphidus arvensis* Starý, *Praon absinthii* Bignell, *Praon dorsale* (Haliday), *Praon exoletum* (Nees), *Paralipsis enervis* (Nees), *Trioxys acalephae* (Marshall), *Trioxys glaber* Starý, *Trioxys parauctus* Starý, *Trioxys pannonicus* Starý, *Trioxys centaureae* (Haliday).

As more typical steppe-species, which do not spread too much in cultured areas, only *Trioxys pannonicus* Starý (a parasite of *Titanosiphon*) and *Lysiphlebus arvicola* Starý (a parasite of *Sipha* spp.) from the above mentioned species may be selected.

As for the endemic species, no such species have been known from the territory of Czechoslovakia till now. Although some species are known only from this country it is believed, due to the occurrence of their hosts, they will be found in neighbouring countries too.

VI.

Bionomics
and
ecology

Development

The laid egg grows several times inside the host and inside it the embryo develops. Then the larva hatches and feeds gradually on various tissues and organs of the host aphid. The less important organs are consumed at first by the larva that at last kills the host. The larvae of instar I and II are not visible through the aphid skin, the last instar larva gradually becomes more and more visible. The last instar larva, before it kills the aphid, is so great that it stuffs almost entirely all the aphid skin. The killed aphid is unable to stay on the host plant and falls down from the plant. In this way it would be separated from the aphid colony and so the parasite larva. On the other hand, the mechanical injury or attack by various predators is more probable on the ground for the parasite larva. For this reason the larva cuts a hole on the ventral side of the aphid skin in the period of killing the aphid and by the secretion of salivary glands it mounts the aphid skin to the surface of the plant. Thus, it lasts in the aphid colony and in this way the contact of the newly emerged parasite with the host is more probable. Nevertheless, according to our laboratory observations the larva is able to spin its cocoon also without mounting the aphid skin to the surface. After the mounting of the skin to the surface of the plant the larva spins the cocoon inside the aphid skin. With the gradual spinning of the cocoon the primarily translucent aphid skin, in which the parasite larva may be recognized, becomes coloured in dependence on the parasite species. Usually the mummified aphid is whitish, yellowish to brownish, only seldom it is black. In the species which spin the cocoon under the empty aphid skin, the colour of the cocoon is whitish, yellowish or brownish, seldom translucent with darkened margin.

The greatest part of aphidiids spin the cocoon inside the empty aphid skin. Only in some genera the larva leaves the aphid skin after killing the aphid and spins under it a separate cocoon, the empty aphid skin being mounted on the top. This cocoon may be either homogenous — silky, or it is translucent with a more heavily built margin.

The thickness of cocoons may be dependent on the season in which the larva spins the cocoon or by other factors. The usual cocoons have a normal appearance, dependent on the species. In some cases a special kind of cocoon is produced, distinguishable from the normal cocoons. In this kind of cocoons the diapause state is spent by the parasite. The diapause cocoons are heavily built and resistant against mechanical injury.

The larva, after the spinning of the cocoon, gradually develops in praepupa, pupa and imago. The adult parasite cuts, by circular movements of the mandibles, a circular chink in the dorsal part of the cocoon (i. e. aphid skin), then presses out the inside of this chink and emerges. As a result of its emergence, a circular hole with a circular lid remains in the mummified aphid. Inside the empty mummified aphid then there are about 15 dark corpuscles and the exuvia of the last instar larva.

The newly emerged adult parasite spends a certain time in cleaning itself, the meconium is excreted, etc. Newly emerged adults are easily recognizable by wide intersegmental membranes under which whitish fat bodies, like in the pupa, are easily recognizable. Depending on the environmental conditions, the imago reaches maturity. The older adults may be recognized according to closely related abdominal segments, between which fat bodies are not visible.

Mating begins as soon as the adult reaches maturity, males being more active, running and searching for females.

Mating. The mating behaviour is very typical in the male. It sits on the female, taps it quickly with the antennae and bends the abdomen. If there is a virgin female, contact usually follows, lasting from some seconds to a few minutes. Once mated, a female refuses to be mated for the second time. The mating behaviour of the female is not remarkable; its wings are laid on the abdomen. A male can mate several females, a female is mated only once in all its life.

The eggs are laid immediately after the female reaches maturity, irrespective of whether it is mated or not.

Progeny

The mated female produces both fertilized and unfertilized eggs, so that both males and females occur in its progeny. This behaviour depends on various factors. If a mated female lays eggs too rapidly — in case of long host's absence, a number of eggs may escape fertilizing as the spermathecal muscles are not tetanized. But usually the mated female uses the sperm till exhausted and as a result unfertilized eggs are produced for the last period of a female's life in spite of the fact that it was mated.

An unmated female produces unfertilized eggs from which only males develop (arrhenotoky). This is the usual case with all the *Aphidiidae*.

To a lesser degree the deuterotoky has been observed in some aphidiids too

[*Lysiphlebus fabarum* (Marsh.)]. In this case the female produces only unfertilized eggs, from which females develop but a certain number of males may also occur in the progeny. This type of parthenogenesis is rather suitable for aphid control as the female parasite does not need mating but nevertheless only female progeny is produced that enables high effectiveness of a parasite.

Sex ratio

Because of the above mentioned reason the sex ratio is rather variable and changes also in the mated female during its life. Nevertheless, certain relations as to the given population (e. g. in the laboratory) may be established.

In our experiments, when *Aphidius megourae* was used as a laboratory object, the number of females was always more numerous, approximately 2 ♀♀ to 1 ♂ and the sex ratio varied as follows: 12th January 1963 — 16 males, 40 females. 22nd January 1963 — 14 males, 29 females. This is the case of the arrhenotokous species. In deuterotokous species the percentage of males in progeny is very low, but the factors influencing it in the aphidiids are still unknown.

Reproductive capacity

It is rather variable in various and in the same species. It depends on a number of factors, the most important of them being the temperature, humidity and food of adults.

The following examples may be mentioned here, results of the research of life histories of two species of aphidiids in the laboratory:

Aphidius ervi Haliday. Under 18—24°C the average fecundity was 50 parasitised aphids per one female parasite, varying from 7 to 77 aphids.

Aphidius megourae Starý. Under 18—24°C the average fecundity was 115 aphids per one female, varying from 5 to 299. In the same species, but under 10—14°C the average fecundity was 20 parasitised aphids per one female parasite, varying from 5 to 51.

Results of our experiments show that the reproductive capacity depends on conditions of adult life. An adult is most reproductive under optimal conditions, while the prolonged life by a lower temperature does not have a positive influence on the reproductive capacity of the parasite. In the above mentioned species — *Aphidius megourae* Starý, the average adult longevity under 18—24°C was 12 days and in 10—14°C 22 days, while the reproductive capacity was higher in the first case and lower in the second one.

98

When a female of an aphidiid is dissected a certain number of mature eggs and a big quantity of immature eggs may be found in its ovaries. The aphidiids, therefore, belong to the synovigenous species.

There is always a big difference between the number of ovarian eggs and eggs that were actually laid. The number of ovarian eggs is always much higher than the actual number of laid eggs.

In *Aphidius megourae* Starý we found more than 100 mature eggs and a big number of unmature eggs in each ovary of a female, while the average number of parasitised aphids was 115 per one female.

Longevity of adults

It is variable in various and in the same species. It depends on three main factors, i.e. temperature, humidity and food of adults. In some cases the longevity of adults is different in males and in females.

Table 1.

Influence of food on the longevity of *Aphidius ervi* Hal. adults. 24°C, 18 hours day-light (after Starý, 1962)

Food	Longevity in days
Without water and food	3
Glucose	3
Water	3
Dissected aphids	3
Dried yeasts	3
Dried yeasts + water	3
Dried yeasts + honey	6—7
Honey + water	6—7
Honey	6—7
Glucose + honey + water	6—7
Dried yeasts + honey + water	6—7
Dried yeasts + water + glucose	9—11
Glucose + water	9—11

Observations on the longevity of adults were carried out in detail in the species *Aphidius megourae* Starý. It was ascertained that the adults are rather resistant also to relatively low temperatures. The newly emerged adults were transferred from breeding of 18—24°C at first to 10—14°C and then to a lower (under 0°C) temperature. The adults responded to low temperatures by torpidity, being unable to sit on the walls of Petri-dishes and falling to the bottom. Temperature under $-5°C$ was lethal. The experiments have shown that in the changing temperature of $+5°C$ to $-5°C$ the adults, having been transferred to $+10°C$ and then to $18-24°C$ can attack aphids in about 15 minutes. A changing temperature of this kind influences negatively the longevity of adults, but they can still reproduce, having even spent a certain time under low temperature conditions.

Table 2.

Longevity of *Aphidius megourae* Starý adults. Food: honey and water. 18 hours day-light, Petri dishes (after Starý, 1964)

Temperature °C		Longevity in days		
		Min.	Max.	∅
10	♀	16	28	21·5
	♂	24	35	29
14	♀	7	26	18
	♂	10	32	24·5
17	♀	6	16	13·5
	♂	7	17	14
21	♀	8	15	15
	♂	7	17	13
24	♀	6	10	8·5
	♂	5	11	8
27	♀	8	11	11
	♂	6	9	9
31	♀	5	6	6·5
	♂	5	7	6

The females used in experiments attacked aphids and laid eggs, after having spent 14 days in $+5°C$ to $-5°C$, in $18-24°C$; in the average 20 eggs were laid and the progeny was normal, including both males and females. It is evident that both eggs and sperma, also in the female's spermatheca, can survive temperature to $-5°C$.

Similar conditions, no doubt, specifically influenced, are believed to exist in other aphidiids too. The above mentioned possibility of adults surviving a lower temperature is important in autumn especially, when the parasites attack e.g. the sexuales of dioecious aphids. Similarly, the ability of parasite adults to spend successfully a relatively long period in a lower temperature is important for their cold storage possibilities in mass-rearing for biological control purposes.

The responses of adults to high temperature were not examined in detail. In these temperatures the mobility and all responses of adults are rather quick and the longevity is shorter.

Food of adults

The problem of food of adults in aphidiids has not yet been successfully solved. As the most suitable laboratory food of adults honey and water was used. Other

Table 3.

Food of *Aphidius megourae* Starý adults. 21°C, 18 hours day-light, Petri dishes, laboratory (after Starý, 1964)

Food	o	Longevity in days
Water	♀	3·5
	♂	4
Water + dried glucose	♀	3
	♂	4
Water + glucose solution	♀	5
	♂	5
Water + honey	♀	15
	♂	13
Water + pollen	♀	8
	♂	9·5
Water + honey + pollen	♀	14
	♂	10
Water + honey-dew of *Megoura viciae*	♀	4·5
	♂	3
Water + honey-dew of *Acyrth. pisum*	♀	3·5
	♂	4
Water + honey-dew of *Aphis fabae*	♀	3
	♂	4

laboratory combinations of food, using combinations of water, dry glucose, glucose solution, honey, pollen, honey dew of aphids, etc. were examined and supported only the suitability of honey as the most suitable laboratory food of aphidiid adults.

There remains a question of the food of adults in the field, as the adult aphidiids are generally supposed to feed on the honey dew of aphids. Although we observed several times some aphidiid species feeding on honey dew of aphids [*Lysiphlebus fabarum* (Marsh.)], the laboratory experiments using honey dew as laboratory food were not successful, although the honey dew of host aphids and other aphids was used.

Judging from the composition of honey dew this kind of food may be theoretically supposed to be suitable.

The newly emerged parasites do not need either food or water for a certain period after their emergence. Females may infest aphids and males may mate females. Nevertheless, after a short period of oviposition the females stop oviposition (after exhausting the primary reflexes of oviposition) and search intensively for water. Mere water as food keeps the female life on the minimal rate that enables a certain number of eggs to be laid. For reaching the optimal rate of adult life a certain kind of adult food is necessary.

Effect of temperature on the rate of development

Like in other insects also in the aphidiids a certain optimal developmental temperature may be experimentally recognized.

The optimal temperature depends on the species. In general, it seems to be lower in the forest-type than in the steppe-type habitats inhabiting parasite species.

Table 4.

Effect of temperature on the rate of development of *Aphidius megourae* Starý (after Starý, 1964)

Temperature °C	Development in days	
	infestation-mummification	mummification-emergence
34	7	— (died)
31	5	5
27	5	5
24	6	5
21	8	6
17	10	8
14	15	12
10	28	17

Under a certain lower temperature the parasite does not develop, while higher temperatures kill the parasite. In some cases the diapause may be also induced experimentally if the parasite is transferred to a high but not lethal temperature. This is also the response of some parasites to high temperature in the field.

Seasonal occurrence. Hibernation. Diapause

The aphidiid adults may be found in general during all seasons in nature from the beginning of May to about the end. of September or beginning of October in Czechoslovakia. Their occurrence is specifical and it is in close relation with the requirements of temperature, with host-aphid life cycle and degree of secondary parasitism.

Differences in the occurrence of the same species may be found in Czechoslovakia in dependence on the districts as, besides the deciduous forest zone, a part of Czechoslovakia belongs to the steppe zone where both the aphids and parasites occur earlier when compared with other districts.

In many species of parasites it seems that they occur some days later in spring than aphids so that in many aphids the first generation (fundatrices) is rarely or never attacked by the parasites and the aphids can reach a certain population level before the first specimens of parasite emerge.

Some species, especially strongly specialized parasites, are usually very closely dependent on the occurrence of their hosts. The adaptation of less specialized species, on the contrary, enables their occurrence in not such a restricted period as they are able to infest quite a number of aphid species. Examples of the mentioned adaptation are mentioned in chapter IV.

Hibernation. In the autumn we may find on dry herbs some mummified aphids, such as various *Dactynotus* and *Macrosiphoniella* spp. on *Centaurea* and *Artemisia*, etc. and others, that contain praepupal stage of aphidiids but from these adults do not develop in this period. To recognize this, experiments on *Aphidius megourae* Starý, the rather suitable laboratory object, were undertaken to study the mode of hibernation of aphidiids in nature. Plants with mummified aphids were transferred from the laboratory breeding at the beginning of November at first to +10°C—14°C and then to natural conditions in a silon cage that was left in a garden throughout the winter. The silon cage was covered with snow, wet if the temperature was above 0°C. The parasite larvae spun cocoons but remained in the praepupal stage, and the mummified aphids were on the leaves till frost came, when all leaves fall. At the beginning of spring of the next year a certain number of mummified aphids was dissected in the laboratory. Inside the mummified aphids living praepupae were found, easily distinguishable from the usual praepupae by their remarkably yellow coloration. A part of the hibernating material was then transferred to the laboratory, at first to $+10°C$ and later to $18-24°C$ where the adults emerged after about two weeks.

This experiment, in connection with our experiment on the longevity of adults, shows that the praepupal stage is the only stage in which the parasite is able to spend the winter period as the adult cannot survive either a low temperature, or the changeable temperatures in spring and in autumn; the latter stimulate or retard its activity that has negative influence on its longevity too.

There is probably one exception, that of *Paralipsis enervis* (Nees) which was found to hibernate in adult stage to an ant-nest. This is probably due to the adaptation of this species, which is a parasite of underground aphids attended by ants, which enables the hibernation of the adult parasite as a member of the ant-nest community. The winter temperature conditions in an ant-nest are suitable both

Table 5.

The occurrence of diapause and non-diapause cocoons in *Ephedrus persicae* Frog.

Host aphid	Plant	Number of samples	%	Diapause cocoons	Period occurrence/month	non-diapause cocoons (+)	diapause cocoons (*)
					V	VII	VII
Allocotaphis quaestionis (CB.)	*Malus silvestris*	1	2.1	+	+*		
Aphis fabae Scop.	*Euonymus europaea*	1	2·1			+	
Aphis idaei v. d. G.	*Rubus idaeus*	1	2·1	+		+*	
Brachycaudus helichrysi (Kalt.)	*Prunus persica*	1	2·1				+
Brachycaudus helichrysi (Kalt.)	*Anthemis* sp.	1	2·1			+	
Brachycaudus helichrysi (Kalt.)	*Melandrium* sp.	2	4·2		+		
Brachycaudus sp.	*Prunus domestica*	3	6·3		+		
Dysaphis devecta (Walk.)	*Malus silvestris*	4	8·3	+	+*		+*
Dysaphis sp.	*Crataegus oxyacantha*	1	2·1		+		
Dysaphis sp.	*Pirus communis*	1	2·1		+		
Dysaphis sp.	*Malus silvestris*	6	12·5	+	+*	+*	+
Dysaphis sp.	*Sorbus torminalis*	2	4·2	+	+*		
Dysaphis sp.	*Sorbus aucuparia*	4	8·3	+	+*	+*	
Myzodes ligustri (Mosl.)	*Ligustrum vulgare*	1	2·1			+	
Myzus cerasi (F.)	*Prunus avium*	10	21	+	+*	+*	+
Myzus cerasi (F.)	*Prunus cerasus*	1	2·1			+	
Phorodon humuli Schrk.	*Prunus domestica*	1	2·1			+	
Rhopalosiphum padi (L.)	*Padus racemosa*	2	4·2	+	+*		
Roepkea marchali (CB.)	*Prunus mahaleb*	1	2·1	+	+*		
Dysaphis sp.	*Pirus communis*	4	8·3			+	
	Total	48					

for ants and other members of the ant-nest community to which the parasite had adapted itself.

Diapause. The diapause in the aphidiids may be understood as an adaptation of parasites to the period that is unsuitable for parasite existence, either if the host is absent in the given type of habitat or if temperature conditions are unsuitable.

The common case of diapause may be found in *Ephedrus persicae* Frog., a parasite of quite a number of leaf- curling aphids in Czechoslovakia, which was studied in detail. As our studies have shown, the mentioned species occurs in Czechoslovakia from May to July. For this reason research was undertaken to show the cause of such an occurrence. It was ascertained that *Ephedrus persicae* Frog. occurs in habitats of the forest type almost exclusively and it is a parasite of mostly dioecious aphids on the primary host plants, where they cause leaf-curling, galls, etc. During several years two types of cocoons i.e. of mummified aphids were observed in this parasite. Further studies have shown that they represent diapause and non-diapause cocoons. The experiments made with diapause cocoons have shown that under field conditions the parasite does not emerge during the rest of the given year but remains inside the cocoon in the praepupal stage. It emerges however in the following year, in about the same period as in the preceding year.

These facts allow us to understand the diapause in this case as an adaptation of the parasite to the host-life cycle as the result of the synchronization of parasite and host development: The parasite falls in the diapause for the whole period till the time the host appears in the same habitat again in the following year.

Another example, but of a little different kind, was recorded from Italy (*Aphidius avenae* Hal.). In this case, the parasite spends in diapause the hot unsuitable summer period where there is a lack of aphids, and appears in the autumn again. As the same species occurs in Czechoslovakia too, a similar adaptation seems to exist also in this country.

VII.

Host
and
parasite
interrelationship

Oviposition

Preoviposition period. When searching for the host aphid the aphidiid female uses its antennae in the primary orientation (macro-orientation). The antennae are held forward, slightly bent downwards. If the aphid is tapped, the antennae are held upward and the oviposition positions follow. Visual orientation is believed to be of no apparent significance as the parasite females run around the aphid in close proximity without paying attention to it until they tap it by the antennae. In other cases, parasite females were observed to try to infest aphid mummies or another female, too.

The intensity of orientation and quickness of tapping etc. depend closely on temperature conditions, rate of contact with the host aphids, etc. In the case of a long rate of the host's absence, the oviposition behaviour is quite untypical when compared with the average behaviour in the presence of a sufficient number of host aphids.

The period between the preoviposition period (i. e. macro-orientation) and the true oviposition (micro-orientation and attack) is different in different species. In parasites of quickly movable and running aphid species, it is rather short, in parasites of almost unmovable aphids it is longer.

Oviposition behaviour. The awaiting position, after the tapping of aphid, is as follows: The female stands on erected legs, the abdomen bent downwards. The true oviposition position is similar except that the abdomen bent is stretched forward, moving until the host is tapped by the end of the ovipositor sheaths. In this phase belongs also the keeping of an aphid by ventral prongs in *Trioxys*, etc. (Plate XII, Fig. 9). Then the sting and oviposition follow, being of a different rate. In some species, parasites of aphids that have strong escape reactions, it is rather quick — about 1 second (*Aphidius ervi* Hal.) or less (*Aphidius megourae* Starý). In other species, parasites of slowly moving aphids, it is rather long — about 40 seconds [*Lysiphlebus fabarum* (Marshall)].

The aphid is stuck in different parts of the body, depending on the parasite species.

106

In some species the closest part of the aphid body is attacked. In this case the legs are often oviposited in which the eggs do not develop. But the abdomen is mostly pricked.

In some species it seems that the parasite female attacks the aphid in a certain part of the body only, placing its eggs in a certain body part of aphid that seems to be in close connection with the later larval development.

The direction of stinging depends on the species, on the shape of ovipositor sheaths and ovipositor. In some species they are almost straight, or curved downward or upward.

If the aphid changes its position after being tapped by the parasite female antennae and before it was pricked the female moves its abdomen searching for the aphid and only afterwards the macro-orientation sets in again. In other cases the female follows the running aphid with abdomen bent forward and tries to prick, using micro-orientation by the setae on the ovipositor sheaths only.

The oviposition of a parasite female may be interrupted or not by an external mechanical stimulus. The response of parasites on the stimuli are specifical. In general, parasites of quickly moving aphids are more sensitive and the parasites of slowly moving aphids less sensitive to mechanical stimuli. The presence of ants has a certain significance too (see chapter IX). The sensitive species break off the oviposition and fly or run off. Disturbing stimuli, therefore, in a number of species decrease successful oviposition. They are particularly active when there is too big a population density in a restricted space, e. g. in a laboratory rearing cage.

After a series of stings a resting period occurs, when the female cleans itself, feeds, or sits motionless on the plant.

Number of eggs per one insertion of ovipositor. Our experiments, in which once infested aphids were removed to prevent the second insertion and later dissected, have shown that eggs are laid nearly at each insertion of the ovipositor, one egg being laid at one insertion. The average behaviour must be, however, classified only.

Host specificity

Host specificity development

Aphids represent a group that developed primarily in forests. The initial host plants were trees. The primary climate zone of aphid occurrence is believed to be the temperate climate zone. The further direction of development of aphids has been connected with the transition from trees and shrubs to herbs and with adaptation for their existence in conditions of a dry climate generally. The contemporary distribution of aphids in general is in temperate and subtropic climate zones, in the tropics only few specialized groups originated. In the mentioned zones aphids inhabit the most different types — humid to arid — of habitats.

The *Aphidiidae* originated from the Ichneumonoid complex of the *Hymenoptera*. The recent family *Braconidae* (as a subfamily of which the *Aphidiidae* are again today mostly classified) is connected as to its origin with the tropics. Some groups of this family differentiated and developed but clearly in a temperate climate zone, where there is also their recent centre of development, e. g. the exodont group of *Alysiinae* that includes parasites of *Diptera,* or the aphidiid wasps that include aphid parasites exclusively.

The phylogeny of host specificity development in the *Aphidiidae* may be classified with difficulty only. We know today only some developmental directions or relations, but they represent group-relations mostly so that the developmental relations between the host and parasite in all the group of aphidiid wasps are only fragmentary. The rule of phylogenetic parallelism may be accepted only in some cases as it does not take the most important ecological factors into account, as is mentioned in some chapters of this book.

Clear cases of phylogenetical relations of host and parasite may be found in the aphid family *Lachnidae* and its parasites. The parasites are specialized exclusively on this aphid group, occurring in forest type habitats (mostly coniferous). Similar examples may be found in some other tropical groups of aphids.

From the greatest part, species of various aphidiid genera parasitize so many aphid groups that the recent generic classification of parasites in relation to aphid groups is hardly possible. Usually the same aphid group is parasitized by various species of different genera in various geographical areas, and vikariant species occur, etc. Nevertheless, it is possible today to classify some typical specific parasite complexes of some aphid groups. In this relationship the different degree of influence of phylogeny and ecology may be recognized.

Host specificity phases

Host habitat finding. As our observations have shown the aphidiid wasps are connected with a certain type of habitat which they primarily seek. According to the habitats we may recognize parasites occurring in forest, steppe type habitats, etc. The degree of their dependence on the host occurrence is various in different species. It depends on the width of host range of the given parasite species. In monophagous species the primary influence of habitat and its preference might be under discussion but in widerly specialized species it is clear if a great number of samples is at hand. In the tables there are mentioned two examples chosen at random of relation to the habitat on the one hand — of a taxonomic unit (genus *Ephedrus*) and on the other hand — of an aphid parasite complex (*Aphis fabae* and its parasites).

Host finding. In the given type of habitat the parasite female finds a host-aphid by macro-orientation, using its antennae (stimuli) by which the host is finally tapped. The micro-orientation, by which various setae on ovipositor sheaths are used or probably setae on accessory prongs too, then may or may not follow in dependence on the host acceptance.

The aphidiids spread in a habitat on longer distances passively by wind. It seems they fly actively on shorter distances as may be recognized from their slow and gradual spread on cultured areas (field) from the neighbouring habitats. This fact was mentioned by various authors and may be supported by our observations too. It seems that arboricolous species of aphidiids are more active as to the flying when searching for aphid colonies. On an aphid host plant and in aphid colonies

Table 6.

The occurrence of various *Ephedrus*-species in various kinds of habitats in Czechoslovakia

Habitat		*E. brevis*	*E. campestris*	*E. cerasiola*	*E. lacertosus*	*E. minor*	*E. nacheri*	*E. persicae*	*E. plagiator*	*E. validus*
Coniferous woods	%								12	
Deciduous woods	%	+							4	
Mixed woods	%				+	+		3	7	+
Parks, orchards, gardens	%			+	+	+	+	70·5	47·5	
Field groves, shrubs, wood-steppe	%		8·7				+	18·5	10·5	+
Meadows, pastures	%		34				+	4	6	
Steppe	%		28·6							
Fields	%		11·6				+		12	
Fallow lands	%		17·1					+	1	
Total No. of samples		1	35	6	14	9	12	68	159	2

Table 7.

The occurrence of parasites of *Aphis fabae* Scop. in various kinds of habitats

Habitat		*Ephedrus plagiator*	*Trioxys angelicae*	*Praon abjectum*	*Lipolexis gracilis*	*Lysiphlebus fabarum*	*Ephedrus persicae*	*Lysiphlebus ambiguus*	*Trioxys acalephae*
Forest type	%	90·6	92	96·4		0·2	+		
Steppe type	%	9·4	8	3·6	+	99·8		+	+
Total No. of samples		32	86	28	6	106	2	1	3

the parasite mostly climbs only. The relations between flying and climbing depend on the given species. Quickly moving species (like *Aphidius megourae* Starý, *Aphidius ervi* Haliday, *Pauesia* spp., etc.) which have strongly developed escape reactions on mechanical and other stimuli in hot weather, use flying in a greater degree than slowly moving species like *Lysiphlebus fabarum* (Marshall) that almost exclusively climbs.

In the case of *Paralipsis enervis* (Nees) there seem to be two phases, connected with the host finding. *Paralipsis enervis* (Nees), being a parasite of underground aphids, searches for aphids in ant nests, where it comes in contact with aphid attending ants. The primary mode of the spread of *Paralipsis enervis* (Nees) is flying. In an ant nest its wings are mutilated by ants and the parasite female cannot spread more using wings but may climb only, its wide spread possibilities being restricted in this way.

Host acceptance. The host acceptance is a complex process and it depends on a number of various factors:

A. Defensory reactions and modifications of the host. They may be a) active (movements of legs, of aphids falling down from the plant after being tapped by the female). In aphid species that use movements of legs as mode of defence against parasite female attack this defence is different in various instars. Lower instars — first and second — are less active and for this reason probably they are more infested by ovipositing parasite females, while in a number of species movements of legs used in higher instar and adult aphids either repel the female or it oviposits in an aphid leg as in a closer aphid body part, so that in both mentioned cases the aphid is in fact unparasitized, as eggs deposited in legs do not further develop. It seems that in certain species, depending on the rate of their development, the movements of legs that repel the parasite are the stimulus for the parasite not to oviposit in the aphid, as due to the rate of development, the larvae of parasites that hatch from eggs deposited in adult aphids, cannot finish the development due to the rate of adult aphid life, which is shorter than the rate of parasite development. This is true in a number of species we studied (*Aphidius ervi* Hal., *Aphidius megourae* Starý, etc.), while in other cases it seems [*Praon abjectum* (Hal.), *Dyscritulus planiceps* (Marshall)] that the rate of adult aphid life is long enough for the parasite development, as mostly alate aphid adults are found to be mummified. This feature is also rather important for the possibilities of aphid spread. More detailed research will show the further dependence occurring in these relations. b) passive. In many aphid species strong wax covers are developed that repel many parasite species. This may be seen from the host — parasite relations in a quantity of parasite species that attack the given host aphid. The production of a quantity of honey dew seems to be also a factor influencing the parasitization degree. From this point of view all galls, leaf-curling, etc. may be classified as an adaptation against natural enemies too as such aphids are attacked by specialized i. e. adapted parasite species. The

110

sclerotization of cuticle may be also an eliminating factor, as it may be found to be apparent from the comparison of ant attended aphids and non-attended aphids.

B. Host movability. This may be classified from two different viewpoints. On the one hand, such aphids are less movable than hungry aphids that search for food. This may be classified as an extraordinary case from the point of view of host × × parasite relationship. On the other hand, if only such aphids are dealt with, the following may be recognized in some parasite species: moving by legs or antennae

Table 8.

Aphidius megourae Starý and *Megoura viciae* Bckt. relationship. Effectiveness and host-instar preference in 10—14°C (after Starý, 1964)

Aphid instar	♀ parasite No.										Total number
	1	2	3	4	5	6	7	8	9	10	
III	—	—	2	2	—	—	1	—	—	4	9
IV	15	—	36	25	—	—	2	7	18	30	133
Adult	14	—	13	4	—	—	2	3	8	13	57
Total number	29	—	51	31	—	—	1	10	26	47	199 ──── 20 per ♀

Table 9.

Aphidius megourae Starý and *Megoura viciae* Bckt. relationship. Effectiveness and host instar preference in 18—24°C (after Starý, 1964)

Aphid instar		♀ parasite No.										Total number %
		1	2	3	4	5	6	7	8	9	10	
III	No.	1	8	—	—	—	5	3	5	1	16	40
	%	20	5	—	—	1·2	3.5	2·2	4·4	1·8	5·4	3·4
IV	No.	2	111	66	70	66	112	93	92	36	231	879
	%	40	75	77	88	79·8	75	67·9	81·6	63	77	75·8
Adult	No.	2	30	20	10	16	32	41	16	20	52	239
	%	40	20	23	12	19	21·5	29·5	14	35·2	17·6	20·6
Total number		5	149	86	80	83	149	137	113	57	299	1158
												∅ about 115 aphids per ♀

aphids slightly are preferred, while entirely motionless sitting and sucking aphids are ignored by the parasite female after macro-orientation i. e. tapping of antennae. Or, in other species, the parasite female oviposits in an aphid of suitable instar irrespective of its movements, etc.

C. Host instar degree. As our observations made on various species of various genera *(Trioxys, Praon, Ephedrus, Lysiphlebus, Aphidius)* have shown, any parasite species prefers a certain host instar that is for a certain reason most suitable for the attack.

The host instar preference must be studied in more or less natural conditions. If, after a longer host absence, whatever instar aphid is added in a Petri-dish that contains a parasite female in the laboratory, such a female oviposits readily in this aphid without preference − as there the possibility of preference does not exist. In such a case the parasite females try also to oviposit in full or empty aphid mummies, in other females, etc. This feature is believed to be caused by a too excited and strong oviposition stimuli. When the "results" obtained by the mentioned method are compared with those obtained in a more or less natural environment where the parasite has the possibility of preference, a big difference is found.

D. The ability of a parasite female to distinguish between parasitised and non-parasitized aphids. Usually, only a host aphid that contains third instar and higher developmental stages of the parasite are ignored by ovipositing females in natural conditions, i. e. in case of preference possibility. This feature partly prevents the superparasitism, as in such a mentioned case the hatched larva is unable to compete successfully at all with higher instar larva. This is also the case of relations among lower instar larvae, too (see superparasitism), but in case of competition between two first instar larvae of different species (multiparasitism) a certain possibility of successful competition may occur.

Host suitability. It depends on the greatest part on the immunity of the host, which depends on host and parasite ecological characteristics, etc.

Factors influencing the host specificity

Habitat. Our observations have shown that the habitat is of the greatest significance in the host specificity of aphidiids. In parasites of monoecious species the importance of relation to the habitat is less distinct as these parasites mostly occur, similarly as their hosts, throughout all the season in the same type of habitat (forest, steppe).

Examples: *Pauesia* spp. − parasites of *Cinara* spp. in forest type habitats. *Diaeretus leucopterus* (Hal.) − parasite of *Protolachnus* spp., *Pauesia unilachni* (Gahan) − parasite of *Schizolachnus* spp., both occurring in forest type habitats. *Aphidius ervi* Haliday and *Praon volucre* (Haliday), parasites of *Acyrthosiphon pisum*, occurring in steppe type habitats.

112

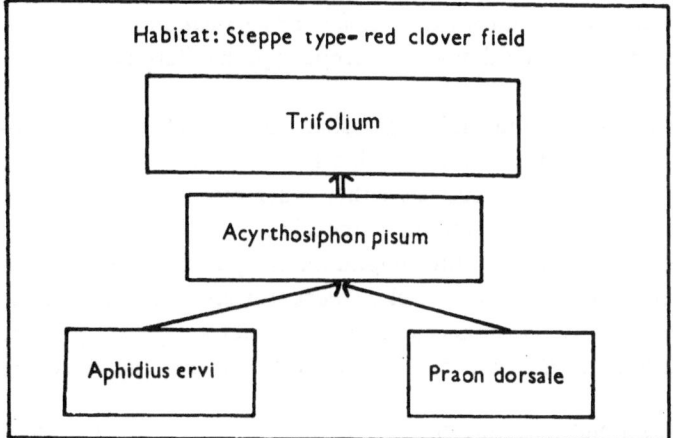

Fig. 4. A scheme of a monoecious aphid species and its parasitization depending on habitat (*Acyrthosiphon pisum* Harris, red clover, parasitized by *Aphidius ervi* Hal., and *Praon dorsale* Hal.).

The relation to the habitat is more apparent in parasites of dioecious aphids, which during their life in connection with the aphid migration from primary to secondary host plants change their habitat throughout the season (forest – steppe – forest). In this case one dioecious aphid species is attacked by different parasite complexes depending on type of habitat in which it occurs.

Example: *Aphis fabae* Scop. – In forest type habitat on primary host plants it is attacked by *Ephedrus plagiator* (Nees), *Trioxys angelicae* (Hal.), *Praon abjectum* (Hal.), in steppe type habitat on secondary host plants it is attacked by *Lysiphlebus fabarum* (Marshall) and *Lipolexis gracilis* Förster. *Brachycaudus* spp. – In forest type habitats on primary host plants attacked by *Ephedrus plagiator* (Nees), *Trioxys angelicae* (Haliday), in steppe type habitats attacked by *Lysiphlebus fabarum* (Marshall), *Lipolexis gracilis* Förster, *Paralipsis enervis* (Nees). *Dysaphis* spp. – In forest

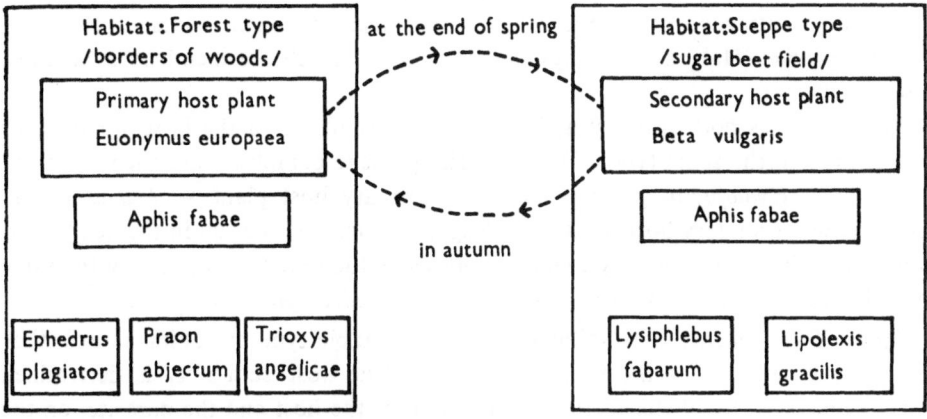

Fig. 5. A scheme of a dioecious aphid species and its parasitization depending on habitat (*Aphis fabae* Scop. on primary and secondary host plants).

type habitats on primary host plants attacked by *Ephedrus persicae* Frog., *Ephedrus plagiator* (Nees), *Trioxys angelicae* (Haliday), in steppe type habitats attacked by *Paralipsis enervis* (Nees).

As far as it is known to us dioecious aphid parasites do not include the monophagous species. After migration of a dioecious aphid host from the given type of habitat the parasites infest other aphid species, either other dioecious species that still occur in the habitat, or some other suitable monoecious species.

Examples: *Trioxys angelicae* (Haliday), after migration of *Aphis fabae* Scop. from forest type habitats at the end of spring, attacks *Rhopalosiphum padi* (L.) that is a dioecious species and namely other monoecious aphid species as *Aphis cognatella* Jones, *Aphis cytisorum* Htg., etc. *Ephedrus plagiator* (Nees) — like in the preceding parasite species.

In other cases dioecious aphid parasites fall in diapause in the period when their dioecious aphid hosts migrate to secondary host plants in another type of habitat.

Example: *Ephedrus persicae* Frog., a parasite of quite a number of leaf-curling aphids (see: Diapause).

Our studies have shown that the type of habitat is more important than if primary or secondary host plant of the aphid occurs in this habitat. In some cases of intermediate habitats it is possible that in forest type habitats (parks) also secondary host plants occur (undergrowth, pathways, etc.) besides primary host plants (shrubs). In this case the aphid parasite complex is mixed. The influence of habitat borders is also important. In this case, usually forest species have greater influence than steppe habitat species.

Example: There occur in parks *Euonymus europaea* shrubs as primary host plants of *Aphis fabae*. There the aphid is infested by three parasite species: *Ephedrus plagiator* (Nees), *Trioxys angelicae* (Hal.) and *Praon abjectum* (Hal.) In a close neighbourhood there is often a flower-bed or a waste place with the secondary host plant — e. g. *Arctium*, etc. On this plant, which occurs in a certain kind of intermediate habitat, the aphid is attacked by steppe species *Lysiphlebus fabarum* (Marshall), and by *Praon abjectum* (Hal.) and *Trioxys angelicae* (Hal.), which are members of a forest community.

In other cases both primary and secondary host plants of the host aphid occur obligatorily in the same type of habitat. The parasite complexes are then identical.

Example: *Hyalopterus pruni* (Geoffr.) — primary host plant — *Prunus* occurs in orchards, secondary host plant — *Phragmites communis* occurs in the neighbourhood of orchards, in pathways, etc. In both cases the aphid is infested by the same parasite species — *Praon volucre* (Haliday) and *Ephedrus plagiator* (Nees).

Host. The existence of a suitable host in a suitable type of habitat is a further necessary condition for the parasite existence. The host suitability is determined on the one hand by phylogenetical relationship of the host and the parasite, on the other hand by the range of the plasticity of parasite-specificity. These two factors are of a different degree of importance and their influence has changed during the

114

phylogeny. We know today both clear cases of strong adaptation of an aphid parasite to the host, and such cases, where the mode of host life is more important than host × × parasite phylogenetical relationship, a number of intermediate cases being known too.

The number of host species present in the given habitat is also rather important and influences the host selection.

Host attending insects. A number of aphid species is attended by ants the relation of which to aphid natural enemies is various. Relation of ants to aphid parasites have often been discussed. Our observations have shown that the aphidiids are quite indifferent to aphid attending ants or more rarely there exists the relation of mutualism between them.

Food of parasite adults. As is known food sources of adult parasites influence also the distribution of parasites. Although our preliminary observation does not show it, it seems that honey dew is the food of the aphidiid adults. Therefore the food of adults exists in the same place as the host aphid.

Ovipositional stimuli of the parasite female. The long-termed lack of host is, no doubt, a strong stimulus for the parasite female. Both in nature and in the laboratory it may cause the attack of a less suitable aphid and the possible increase of the host-list. The act of mating or the presence of sperm in the spermatheca do not have any marked effect on the ovipositional behaviour of the female: Mated or unmated females attack the host in the same degree of intensity.

Aphid host plant. Our field observations have shown that the aphid host plant does not seem to have any great influence in the host selection of aphid parasites. Nevertheless, we observed some probable cases of influence of aphid host plant on the host selection of aphid parasites.

Example: *Trioxys angelicae* (Haliday) is a common parasite of various aphid species in habitats of forest type. *Aphis fabae* occurs here commonly too (the problem of biological races of this species is not dealt with here) on *Euonymus europaea* and *Philadelphus coronarius*. Although the aphid is attacked heavily on *Euonymus europaea*, it is attacked only to a lesser degree on neighbouring shrubs of *Philadelphus coronarius*. In this case, the influence of the aphid host plant is possible.

Geographic distribution. Many examples of different influences of geographic distribution on the parasite specificity might be mentioned. As Czechoslovakia is a small country, where such an influence cannot be apparent, the undermentioned examples were selected to show at least general principles or possibilities of such an influence.

The parasite fauna of the same aphid species often varies in different geographical areas.

Example: *Hyalopterus pruni* is attacked in Europe and in all its distribution area by *Ephedrus plagiator* (Nees) and *Praon volucre* (Hal.), while in South Europe and Central Asia only, it is attacked by *Aphidius transcaspicus* Telenga.

The parasite fauna of the same aphid genus often varies in different geographical areas too.

Example: *Periphyllus* spp. are attacked in Europe by *Aphidius setiger* Mackauer, in the Far East *Periphyllus* spp. are attacked by closely related *Aphidius areolatus* Ashmead.

In other cases, one aphid species is attacked by the same aphid parasite in all its distribution area.

Example: *Brevicoryne brassicae* (L.) is a cosmopolitan aphid pest that is infested in all its area by the parasite *Diaeretiella rapae* (M'Int.).

Schizolachnus spp. are distributed in the Palaearctic region and in a part of the Ethiopian region, where they are attacked by the parasite *Pauesia unilachni* (Gahan).

The same parasite species often prefers different host species in different parts of its distribution area.

Example: *Ephedrus campestris* Starý in Central Europe is a parasite of *Macrosiphoniella* and *Dactynotus* spp. In the Far East it attacks *Megoura viciae* although the same aphid species is distributed in Europe too.

One parasite species sometimes attack various species of the same aphid group (*Ephedrus persicae* Frog.) or of the same genus (*Aphidius absinthii* Marshall).

Some groups of parasites of similar specificity are often vikariant in distribution (genera *Lysiphlebus* and *Lysiphlebia*).

Host and parasite

Terminology

Definitions of parasitism

Parasitocoenosis is an association of organisms that inhabit a certain host. In the case of aphids and their parasites this relationship, representing but a part of the total parasitocoenosis only, is simple and the following cases of relations may occur:

1. Primary parasitism. In this case only the primary parasite larva is present in a host. This is a common case in all the aphidiid parasites, which are the primary parasites of aphids exclusively.

2. Superparasitism. In this case two or more larvae of the primary parasite of the same species are present in a host. Superparasitism may occur commonly in aphid parasite breedings especially if there is a greater population density of parasites, as the ovipositing parasite females do not distinguish between parasitised and non-parasitised aphids.

3. Multiparasitism. In this case larvae of at least two species of primary parasites are present in a host. This case may be met especially where an aphid species is

attacked by several parasite species in a given type of habitat. For example: *Aphis fabae* Scop. is attacked in forest type habitats by three parasite species; if their population density is high, a case of multiparasitism may occur if an aphid is attacked by two ovipositing females of two parasite species as the parasite females do not distinguish aphids which they or another female of the same or another parasite species had attacked.

4. Hyperparasitism I. One primary parasite larva and one secondary parasite larva are present in a host. This is a common case in the aphidiid parasites which are commonly infested by the hyperparasitic chalcids, proctotrupids and cynipids.

5. Hyperparasitism II. Primary, secondary and tertiary parasite larvae are present in a host. The really obligatory tertiary parasites of aphids are not known. Secondary aphid parasites include, however, internal and external species. If a primary parasite larva is attacked by an internal secondary parasite larva, they may be both attacked by an ectoparasitic parasite larva which is then classified as a tertiary parasite.

Definitions of parasitism based on the number of host species attacked

Usually, three types of parasitism, based on the number of host-species attacked, are distinguished: Monophagous, oligophagous and polyphagous.

We found the above mentioned classification of host as quite unsatisfactory for aphid parasites specificity. For this reason we decided to use the classification of certain types in the host specificity of aphid parasites. This classification eminates from the relation of a parasite to a certain host or host group.

Type 1: The parasite specificity is restricted to a single host species ("monophagous").

Examples: *Dyscritulus planiceps* (Marshall) — *Drepanosiphum platanoidis* Schrk. on *Acer pseudoplatanus. Lysiphlebus fritzmuelleri* Mackauer — *Aphis craccae* L. on *Vicia cracca. Trioxys pannonicus* Starý — *Titanosiphon artemisiae* (Koch) on *Artemisia campestris.* Many *Trioxys* spp.

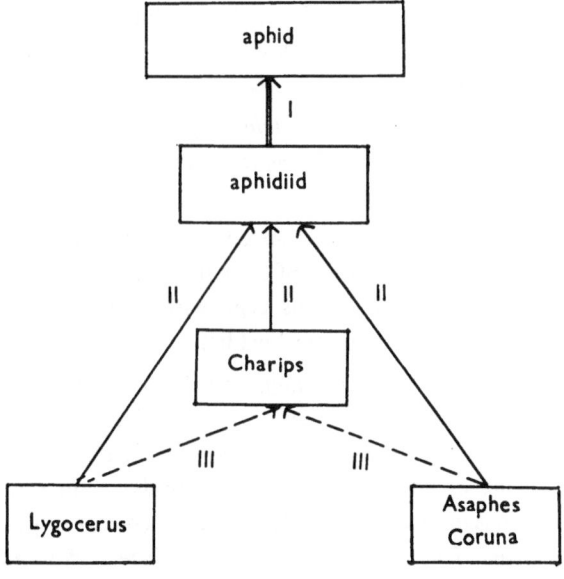

Fig. 6. A scheme of primary, secondary and tertiary parasitism. I — primary, II — secondary, III — tertiary parasites.

117

Type 2: The parasite specificity is restricted to two or more host species of the same genus ("oligophagous").

Examples: *Aphidius absinthii* Marshall — *Macrosiphoniella* spp. *Aphidius funebris* Mackauer — *Dactynotus* spp. *Aphidius pterocommae* Ashmead — *Pterocomma* spp. *Diaeretus leucopterus* (Haliday) — *Protolachnus* spp. *Lysiphlebus salicaphis* Fitch — *Chaitophorus* spp. *Monoctonus angustivalvus* Starý — *Nasonovia* spp. *Monoctonus crepidis* (Haliday) — *Nasonovia* spp. *Praon pubescens* Starý — *Nasonovia* spp. *Aphidius hieraciorum* Starý — *Nasonovia* spp. *Pauesia unilachni* (Gahan) — *Schizolachnus* spp. *Protaphidius wissmannii* (Ratzeburg) — *Stomaphis* spp. *Trioxys acalephae* (Marshall) — *Aphis* spp., etc.

Type 3: The parasite specificity is restricted to two or more genera of the same host group, more or less closely related. Other host groups are not parasitized ("oligophagous").

Examples: *Aphidius salicis* (Haliday) — *Cavariella* spp., *Semiaphis* spp. *Lysaphidus erysimi* Starý — *Lipaphis* spp., *Pseudobrevicoryne* spp. *Ephedrus campestris* Starý — *Macrosiphoniella* spp., *Dactynotus* spp. *Trioxys centaureae* (Haliday) — *Macrosiphoniella* spp., *Dactynotus* spp.

Type 4: The parasite specificity is restricted to two or more genera of the same group of hosts, more or less closely related. Other host groups are rarely parasitized as subsidiary or additional hosts ("oligophagous — polyphagous")

Examples: *Diaeretiella rapae* (M'Intosh) — main host: *Brevicoryne brassicae* L., *Hayhurstia atriplicis* (Geoffr.), *Myzodes persicae* Sulz., subsidiary and additional hosts: *Brachycaudus* sp., *Sitobium* sp.

Type 5: The parasite specificity includes few host genera of the same group to which the main host also belongs, but few other host-groups are often attacked. The mode of host life plays a role ("polyphagous").

Examples: *Trioxys angelicae* (Haliday) — main host: *Aphis* spp., subsidiary and additional hosts: *Rhopalosiphum* sp., *Brachycaudus* sp., *Dysaphis* sp., *Ephedrus nacheri* Quilis — main host: *Hayhurstia* sp., subsidiary host: *Cryptosiphum* sp., both being leaf curling aphids.

Type 6: The parasite specificity includes few or some host genera of various host groups. The mode of host life plays the most important role ("polyphagous"). Nevertheless, also in this type with wide ecological plasticity some aphid groups are lacking in a numerous host list.

Examples: *Ephedrus plagiator* (Nees) — parasite of leaf curling or in dense colonies living aphids (*Acyrthosiphon, Aphis, Brachycaudus, Ceruraphis, Dysaphis, Macrosiphum, Prociphilus, Rhopalosiphum, Schizoneura*, etc.). *Ephedrus persicae* Frog. — parasite of various leaf curling aphids, main hosts — species of the *Anuraphidine* and *Myzine* groups. *Paralipsis enervis* (Nees) — parasite of root aphids (*Aphididae — Anuraphidinae, Thelaxidae — Anoeciinae, Eriosomatidae — Fordinae*).

118

Such a similar classification of host specificity needs a large quantity of reared samples. It will change a little in dependence on data on host specificity of the given parasite species that will be obtained in other areas of its distribution.

Host and parasite relationship in ontogeny

The interrelationship between the host and parasite in ontogeny is a complex relation, the mutual effects on host and parasite being numerous. The main effects, according to our observations, are the following:

1. Reaction of the host to infestation

There are two kinds of defensory reactions and modifications of the aphids:

a) External. There are many aphid species (e. g. *Megoura viciae*) that nearly ignore the presence of the parasite female; only having been pricked by its ovipositor they show a little stimulated activity, moving their abdomen upwards, slightly moving their legs, etc., but mostly they do not take their rostrum out of the plant. Escape reactions of the aforementioned type are more developed in higher instar aphids and adults, which, however, are not preferred by the female.

In other aphid species they almost lack all the responses to the parasite female attack (*Aphis fabae* Scop.). Their response to oviposition of the parasite female is a slight movement, the sucking being uninterrupted. Sometimes, the response depends on the parasite species that attack the aphid: When an aphid of instar I – II is attacked by *Lysiphlebus fabarum* (Marsh.), almost no response is apparent. If such an aphid is attacked by *Trioxys angelicae* (Hal.), whose female uses the prongs of the last sternite to keep the aphid, the attacked aphid produces waxy secretion from the siphunculi.

Other aphid species's response *(Acyrthosiphon pisum)* on the parasite female's attack is the prompt fall of the aphid from the plant so that the aphid escapes from the parasite's oviposition possibilities. Such an escape reaction, which depends on the temperature too, is apparent in big colonies of aphids, where one or few parasite female attacks is enough to spread all the colony, whose members fall from the plant.

b) Internal. The internal reactions are different in individual species. They depend closely on the host and parasite interrelationship. In the case of an adapted parasite, the internal reactions of the host on the presence of parasite egg or larva are practically negligible, and the parasite can develop and finish its development successfully. In the case of an unadapted parasite, the internal reactions of the host on the presence of parasite egg or larva are rather strong, special kinds of cells, etc. are produced and the parasite embryo dies in consequence.

119

2. Effect of parasitization on host life

An aphid containing a higher instar parasite larva moves only slowly or not at all, responding very feebly to external mechanical stimuli. Before their deaths such aphids sometimes climb to more distant parts of a plant, such as the underside of leaves, etc. where they then are killed and mummified by the parasite larva. The mummified aphids are usually dispersed over a plant, single specimens or smaller groups. The dispersion depends on the population density of suitable host instars infested, the climbing and movability of parasitized aphids, parasite population density, mode of parasite oviposition, etc. (Plate X, Figs. 1, 3, 4).

The parasitization retards the host's development when compared with non-parasitized host. In the case of *Megoura viciae* and *Aphidius megourae* relationship the parasitization in instar's I and II delays the host's development if compared with non-parasitized hosts for about the rate of an instar period. Aphids infested in instars I–II reach instar (III)–IV when they are mummified, while at the same time the non-parasitized host reaches the adult stage. The parasitization in higher instars (III, IV) does not delay the host's development, but becomes evident in the shortening of the host's reproductive period and in the decreased number of host progeny (Tab. 7).

The parasitization has also an influence on the rate of host reproduction. This

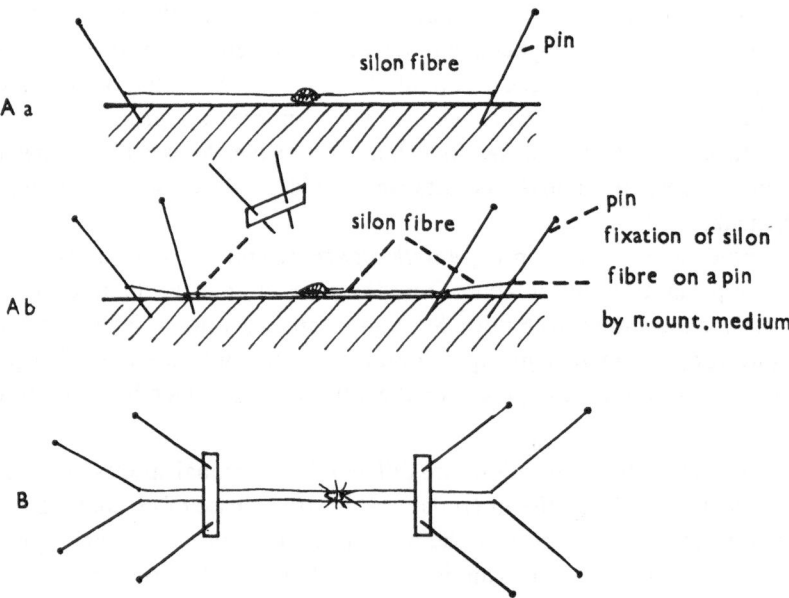

Fig. 7. Aphid dissection technique. Aa — sticking of pins with mounted silon fibres for aphid fixation. Ab — adjustment of fibres along the aphid, across its legs; fixations by plastic plates. B — seen from above.

depends on the instar in which the aphid was attacked *(Aphidius megourae × Megoura viciae)*:

In the first few days after the infestation there are no differences between the parasitized and non-parasitized aphids; in the following days the changes depend on the instar in which the host was infested.

A host infested in instar I is mummified in instar III–IV at the time when a non-parasitized host reaches maturity. The parasite emerges at the time when the nonparasitized host has been producing progeny for several days.

A host infested in instar II is mummified in instar IV. At this time the non-parasitized aphid reaches maturity. The parasite emerges at the time, when the non-parasitized aphid has been producing progeny for several days.

A host infested in instar III reaches the adult stage and produces its progeny for several days before being mummified. The parasite emerges at the time when the non-parasitized host progeny reaches instar IV.

The host aphid infested in instar IV reaches maturity and produces its progeny for several days. The parasite emerges at the time when the non-parasitized host progeny reaches maturity.

Table 10.

Aphidius megourae Starý and *Megoura viciae* Bckt. relationship. Effectiveness of parasitization on host reproduction and effects of parasitization on the rate of development of P and NP hosts

		Development in days														
		1	2	3	4	5	6	7	8	9	10	11	12	13	14	15
P	I	I								III-IV-M	----	----	------	----	----	PE
NP	I	I								A	Ar	Ar	Ar	Ar	Ar	......
P	II	II								IV—M	----	----	------	----	----	PE
NP	II	II								A	Ar	Ar	Ar	Ar	Ar	......
P	III	III			A	Ar	Ar	A—M	----------	----	----	------	----	----	PE
NP	III	III			A	Ar	Ar	Ar		Ar	Ar	----	------	----	----	F1/IV
P	IV	IV			A	Ar	Ar	Ar	Ar	A—M	----	----	------	----	----	PE
NP	IV	IV			A	Ar	Ar	Ar	Ar	----------	----	----	F1/IV			
P	A	A	Ar	Ar	Ar	----	----	----	------							
NP	A	A	Ar	Ar	Ar	----	----	----	------							

I, II, III, IV, A — host instars, M — mummification of aphid by parasite in the given instar, Ar — adult host reproduction, P — parasitized host, NP — nonparasitized hosts, PE — parasite adult emergence, F_1/IV, A — first generation of the host (after Starý, 1964).

121

The host aphid infested in the adult stage produces progeny in the usual way and is not mummified, as it dies before the parasite is able to finish its development.

The form of progeny in aphids is determined mostly in the second half of embryonic development and sometimes during lower instars by the aphid female. In the first mentioned case *(Megoura viciae)* the influence of parasitization on the form of host progeny is negligible. In the second case *(Aphis fabae)*, if the aphids are parasitized in lower instars, the development of their wings is deeply modified because of the influence of parasite larva on the host's life.

3. Effects of parasitized host on the parasite

According to our observation the parasitized host has no influence on the sex determination and on the rate of the parasite. The sex of the parasite is determined by the parasite female that produces fertilized or unfertilized eggs. The rate of parasite development does not depend on the instar in which the host was attacked as it is apparent from breeding of parasitized and non-parasitized hosts in the same laboratory conditions.

The host can have a certain influence on the size and fecundity of the parasite *(Aphidius megourae × Megoura viciae)*. Aphids parasitized in instar I are killed and mummified in instar III which is smaller than mummified instar IV and adult aphid. As a consequence, a smaller parasite specimen emerges. This is, however, a laboratory case, as under natural conditions where preference possibilities exist, only the most suitable instar aphids for attack are clearly preferred by the ovipositing parasite female.

Host and parasite relationship in phylogeny

Taxonomic aphid groups and parasites

The purpose of this chapter is to show the complexes of parasites that attack individual taxonomic groups of aphids. In many cases the close specialization of a parasite to host is apparent, on the other hand clearly ecological adaptation may be recognized in many parasite species, many intermediate cases existing too.

The classification of host-specificity of a different parasite species is given in the main features only if necessary. More detailed data are presented in the taxonomic chapter of this book. A more detailed list of parasites of the aphid species is given in host × parasite catalogue.

For the general system of aphids the classification of Börner, with some changes, has been followed.

122

In the list of aphid groups and their parasites only such genera, etc. are mentioned, from which parasites were bred in Czechoslovakia. Literary data have been omitted for this reason. More comprehensive data may be found in revisions of individual genera and ecological papers mentioned in references.

1. Family *Lachnidae*
1. Subfamily *Cinarinae*
1. Tribe *Protolachnini*

Protolachnus-species are attacked by a strictly specialized parasite-complex that is represented by *Diaeretus leucopterus* (Haliday) and *Praon bicolor* Mackauer. Both hosts and parasites occur in forest type (coniferous and mixed) habitats exclusively.

2. Tribe *Schizolachnini*

Schizolachnus-species are parasitized by *Pauesia unilachni* (Gahan), which is a strictly specialized parasite. Both host and parasite occur in forest type habitats (coniferous and mixed woods).

3. Tribe *Cinarini*

Members of this tribe (genera *Cinara, Todolachnus, Buchneria, Cupressobium*) are attacked by various members of the parasite genus *Pauesia*. They occur in forest type habitats (coniferous and mixed woods), being typical representatives of forest communities. The specificity of parasites of *Cinarini* is of a different type. The greatest part of parasites is specialized on *Cinara* spp. More apparent specialization may be recognized in *Todolachnus abieticola,* which is attacked by the specialized parasite *Pauesia grossa* (Fahr.). A similar example is that of the *Cupressobium* — species that are attacked by specialized parasites *Pauesia cupressobii* (Starý) and *Pauesia juniperorum* (Starý).

2. Subfamily *Lachninae*
1. Tribe *Lachnini*

Maculolachnus-species *(M. submacula)* are attacked by the specialized parasite *Pauesia maculolachni* (Starý). Both host and parasite occur in forest type or intermediate habitats.

In the genus *Lachnus* parasites were not recognized although numerous material was reared.

2. Tribe *Stomaphidini*

Stomaphis-species are parasitized by the strictly specialized parasite *Protaphidius wissmannii* (Ratz.). Host and parasite occurs in forest type habitats (deciduous and mixed woods).

3. Subfamily *Traminae*

Parasites were not recognized in Czechoslovakia.

Conclusions: The *Lachnidae* are attacked by rather specialized complexes of parasites. The common and typical feature of these complexes is that their members do not attack members of other aphid groups. These complexes of parasites may be further subdivided into parasites of different subfamilies and tribes: *Protolachnini*, *Schizolachnini* and *Cinarini*, which include the subfamily *Cinarinae*, and *Lachnini*, *Stomaphidini*, which include the subfamily *Lachninae*. All the members of the *Lachnidae* are holocyclic monoecious species that occur during all the season in the given type of habitats, where they are attacked by corresponding parasite complexes. All the parasite species of the *Lachnidae* are also typical members of the forest, to a lesser degree intermediate habitats.

2. Family *Chaitophoridae*

1. Tribe *Periphyllini*

Periphyllus-species are parasitized by the strictly specialized parasites — *Aphidius setiger* Mackauer and *Trioxys falcatus* Mackauer. Both parasites and hosts occur in forest type habitats (deciduous and mixed woods). *Periphyllus*-species are monoecious.

2. Tribe *Chaitophorini*

Chaitophorus-species are attacked by a strictly specialized parasite — *Lysiphlebus salicaphis* (Fitch). Another parasite — *Ephedrus plagiator* (Nees) is less specialized and attacks a number of in forest type habitats living aphids too. Both parasites mentioned are typical for forest (deciduous and mixed woods) habitats. *Chaitophorus*-species are monoecious.

124

3. Tribe *Atheroidini*

Sipha-species are monoecious species, typical members of steppe communities. They are parasitized by specialized species — *Lysiphlebus arvicola* Starý, which occurs in steppe habitats.

Conclusions: The three tribes of the *Chaitophoridae* are parasitized by specialized complexes of parasites, to a lesser degree by less specialized parasites. The tribe *Periphyllini* is attacked by a rather specialized parasite complex that does not parasitize other tribes of the *Chaitophoridae* or other aphid groups. On the other hand, the *Chaitophorini* and *Atheroidini*, although each of them is parasitized by a specialized parasite complex that does not attack other aphid groups, have certain taxonomic relations as to the parasite fauna. The taxonomically closest species may be found in the parasites of *Thelaxes*-species, which could show a certain taxonomic affinity of *Thelaxes* and the above mentioned tribes of the *Chaitophoridae*.

Parasites of *Periphyllini* and *Chaitophorini* are typical members of forest and intermediate types of habitats, while parasites of *Atheroidini* are members of steppe communities.

All the three mentioned tribes of *Chaitophoridae* include holocyclic monoecious species that occur during all the season in the given type of habitat. For this reason they are attacked by corresponding parasites complexes.

3. Family *Callaphididae*

1. Subfamily *Phyllaphidinae*

1. Tribe *Symydobiini*

Symydobius-species are monoecious, occurring on *Betula* spp. in forest type (deciduous, mixed woods) habitats. They are infested by *Trioxys betulae* (Marshall), which is a specialized parasite.

2. Tribe *Phyllaphidini*

Phyllaphis-species *(P. fagi)* are monoecious and occur in forest type habitats (deciduous and mixed woods). There they are parasitized by a strictly specialized species *Trioxys phyllaphidis* Mack.

125

3. Tribe *Drepanosiphonini*

Drepanosiphum-species (*D. platanoidis* Schrk.) are monoecious, typical inhabitants of forest type (deciduous, mixed woods) habitats. They are parasitized by a strictly specialized complex of parasites: *Trioxys cirsii* (Curtis), *Dyscritulus planiceps* (Marshall), *Monoctonus pseudoplatani* (Marshall).

2. Subfamily *Callaphidinae*

Species of the genera *Chromaphis, Eucallipterus, Myzocallis, Tuberculoides* and *Tinocallis* are typical members of forest type (deciduous, mixed woods) communities. They occur as monoecious species on leaves of various trees *(Juglans, Tilia, Corylus, Quercus, Ulmus)*. The greatest part are parasitized by two typical parasites − *Praon flavinode* (Haliday) and *Trioxys pallidus* (Haliday). To a lesser degree they are attacked also by more strictly specialized parasites *(Trioxys hortorum* Starý × *Tinocallis saltans)*, or widely specialized parasites (*Ephedrus plagiator* Nees).

3. Subfamily *Therioaphidinae*

Therioaphis-species are monoecious aphids, inhabiting steppe type habitats where they occur on *Medicago, Melilotus*, etc. There they are infested by a strongly specialized parasite, *Praon exoletum* (Nees) and *Trioxys complanatus* Quilis M. P.

4. Subfamily *Saltusaphidinae*

Parasites have not been bred in Czechoslovakia till now, although some strictly specialized parasite species are known from other European countries.

Conclusions: All the subfamilies of the *Callaphididae* are parasitized by rather specialized parasite complexes. The influence of more widely specialized parasites that attack also other aphid groups is almost negligible. This may be understood from the fact that the *Callaphididae* include monoecious species exclusively, so that the host × parasite development enabled the creation of such strictly specialized groups of parasites. Nevertheless, the influence of habitats is also apparent here as it is recognizable from the comparison of two closely related species of parasites − *Trioxys pallidus* (Haliday) and *Trioxys complanatus* Quilis M. P.

In the greatest number of cases the close specialization of individual parasite species for the host genera had developed. The only exception is in the subfamily *Callaphidinae*, where the host-specificity of parasites covers almost all the genera of the mentioned group.

126

The greatest part of parasites of the *Callaphididae* belong to the genus *Trioxys*. In almost all the species of this genus the degree of specialization is the highest in all the *Aphidiidae* both as to the morphology and host-specificity.

4. Family *Aphididae*

1. Subfamily *Pterocommatinae*

Pterocomma-species are monoecious, occurring in forest type and intermediate habitats (deciduous and mixed woods, shrubs along ditches, parks, etc.). They are infested by a specialized parasite — *Aphidius pterocommae* Ashmead that does not parasitize other aphid groups.

2. Subfamily *Aphidinae*

1. Tribe *Rhopalosiphonini*

Hyalopterus-species (*H. pruni* Geoffr.) are dioecious, occurring in forest and intermediate type of habitats. Both on primary hostplants — *Prunus* and secondary host plants — *Phragmites communis,* they are parasitized by widely specialized parasites *Praon volucre* (Haliday) and *Ephedrus plagiator* (Nees).

Rhopalosiphum-species: The dioecious *R. nymphaeae* is parasitized in steppe type habitats by *Lysiphlebus fabarum* (Marsh.). The dioecious *R. padi* (L.) is attacked in forest type habitats (deciduous woods and parks) on primary host plant by typical parasites that occur in these habitats: *Praon abjectum* (Haliday), *Ephedrus plagiator* (Nees), *Trioxys angelicae* (Haliday).

Schizaphis-species *(S. scirpi)* are attacked in steppe type and intermediate habitats by *Diaeretiella rapae* (M'Int.) which is a typical parasite of some Myzine aphids.

2. Tribe *Aphidini*

Species of the genus *Aphis* include both the monoecious and dioecious species, occurring both in forest and steppe type habitats. Some of the species occur in spring as root aphids, having close association with ants. Due to the life history of aphids, there is a number of parasite complexes of the *Aphis*-species. To a comparatively low degree there are strictly specialized parasites represented: *Lysiphlebus fritzmuelleri* Mackauer, on *Aphis craccae* (L.), *Trioxys glaber* Starý on *Aphis galliiscabri* (Schrk.), *Trioxys genistae* Mackauer on *Aphis genistae* (Scop.), ? *Trioxys macroceratus* Mackauer on *Aphis podagrariae* Schrk. The greatest part of aphids, both mono- and dioecious, is parasitized by complexes of parasites depending on the habitats in which they occur.

The forest type parasite complex represent the following species: *Trioxys angelicae* (Haliday), *Praon abjectum* (Haliday), *Ephedrus plagiator* (Nees), *Lysiphlebus ambiguus* (Haliday).

The steppe parasite complex includes the following species: *Lysiphlebus fabarum* (Marshall), *Trioxys acalephae* (Marshall), *Lipolexis gracilis* Förster. Some species of the mentioned complexes occur also in intermediate habitats too [*Lipolexis gracilis* Först., *Trioxys acalephae* (Marsh.), *Lysiphlebus ambiguus* (Hal.)]

Protaphis-species occur in steppe habitats. They are infested by *Lysiphlebus fabarum* (Marsh.) on the upper part of plants and on roots by *Lysiphlebus dissolutus* (Nees).

3. Tribe *Cryptosiphonini*

Cryptosiphum-species are monoecious, gall-producing aphids, occurring in steppe type habitats. They are parasitized by *Ephedrus nacheri* Quilis M. P., probably due to the mode of host life.

3. Subfamily *Anuraphidinae*

1. Tribe *Acaudini*

Parasites have not been bred in Czechoslovakia

2. Tribe *Anuraphidini*

This tribe includes both the monoecious species that do not change the type of habitat as they occur on trees only, and the dioecious species that change the type of habitat because of the migration from trees and bushes (primary host plants) to herbs (secondary host plants). They occur, therefore, in forest and steppe habitats. The Anuraphidine aphids have typical bionomical features: In forest habitats, on primary host plants, they occur as leaf-curling aphids; in steppe habitats, on the secondary host plants, they occur on herbs — on roots, root-collars, lower or upper parts of plants.

The mentioned features were also very important in the creation of parasite complexes. The species, both mono- and dioecious, which occur in forest type habitats and cause leaf-curling on trees (*Roepkea, Ceruraphis, Allocotaphis, Dysaphis, Brachycaudus*) are parasitized by a special parasite complex, which is adapted to the host life and host × parasite phylogeny relationship. The importance of these two factors is different in individual parasite species. The mentioned complex includes the following species: *Ephedrus persicae* Frog., *Ephedrus plagiator* (Nees),

128

to a lesser degree *Trioxys angelicae* (Haliday). In steppe habitats, a clear difference may be found between the influence of the host and the host life. The aphid species that occur on roots of secondary host plants are parasitized by a typical root aphid parasite *Paralipsis enervis* (Nees). Such aphids, which live in free colonies on plants are attacked by a parasite complex, which includes a typical steppe species: *Lysiphlebus fabarum* (Marsh.), *Lipolexis gracilis* Först., to a lesser degree *Diaeretiella rapae* (M'Int.). More specialized parasites like *Lysiphlebus melandriicola* Starý, parasite of *Brachycaudus lychnidis* on *Melandrium,* and *Ephedrus persicae* Frog., parasite of *Brachycaudus* on secondary host plants, are rare.

4. Subfamily *Myzinae*

Hayhurstia-species [*H. atriplicis* (L.)] are monoecious, leaf-curling aphids, occurring in steppe type habitats. They are parasitized by a typical Myzine aphid parasite — *Diaeretiella rapae* (M'Int.), and by *Ephedrus nacheri* Quilis M. P., which attacks also other leaf-curling aphids *(Cryptosiphum)*.

Brevicoryne-species *(B. brassicae L.)* are monoecious, occurring in steppe type of habitats. They are parasitized by *Diaeretiella rapae* (M'Int.) and *Praon volucre* (Hal.), both typical parasite species that occur in steppe habitats.

Pseudobrevicoryne-species (*P. erysimi* Holman) are monoecious, leaf-curling, in steppe habitats occurring aphids. They are parasitized by the specialized parasite — *Lysaphidus erysimi* Starý.

Lipaphis-species (*L. erysimi* Kalt.) are monoecious, leaf-curling aphids, occurring in steppe habitats. They are attacked by the same parasite species as the *Pseudobrevicoryne*-species.

Semiaphis-species (*S. dauci* F.) are monoecious and occur in steppe habitats. They are attacked by a typical parasite species — *Aphidius salicis* Hal. which attacks some other aphid groups too, and by *Trioxys spinosus* Starý, which is a strictly specialized parasite.

Hyadaphis-species are monoecious or dioecious, occur in forest or steppe type of habitats. Monoecious species (*H. bupleuri* Boerner) and dioecious species (*H. mellifera* Hottes) in steppe habitats are parasitized by *Trioxys brevicornis* (Hal.), which is a typical parasite of this aphid group. Dioecious aphid species (*H. mellifera* Hottes) in forest type habitats are attacked by a typical forest habitat occuring parasite — *Ephedrus persicae* Frog.

Hydaphias-species are typical inhabitants of steppe habitats. They are monoecious, causing deformations of stem-tops. They are attacked by a specialized parasite *Trioxys parauctus* Starý and by a typical parasite of many Myzine aphids — *Aphidius matricariae* Hal.

Staegeriella-species (*S. necopinata* Boerner) are monoecious aphids, occurring in steppe type habitats. They are attacked by *Trioxys brevicornis* (Haliday), which is a typical parasite of this aphid group.

Decorosiphon-species are monoecious. They occur in a special type of habitat — in peat bogs. They are parasitized by a specialized parasite — *Diaeretellus ephippium* (Haliday).

Coloradoa-species are monoecious typical inhabitants of steppe habitats. They are attacked by a specialized parasite — *Lysaphidus arvensis* Starý.

2. Tribe *Myzaphidini*

Myzaphis-species (*M. rosarum*) are monoecious aphids, occurring in intermediate and forest habitats. They are attacked by a specialized parasite — *Ephedrus minor* Stelfox.

Passerinia-species (*P. tetrarhoda* Walk.) have the main features and parasites the same as in *Myzaphis*.

3. Tribe *Liosomaphidini*

Liosomaphis-species occur in forest type and intermediate habitats. They are monoecious. Their parasites are specifical — *L. abietinum* (Walk.) occurs in coniferous forests and is attacked by *Lysaphidus schimitscheki* Starý. *L. berberidis* (Kalt.) is attacked by a specialized parasite *Aphidius hortensis* Marshall, to a lesser degree by the widely specialized *Ephedrus plagiator* (Nees).

Cavariella-species are dioecious, occurring in forest and intermediate habitats. On primary host plants they are attacked by *Aphidius salicis* Hal., on secondary host plants by *Aphidius salicis* Hal., *Ephedrus minor* Stelf. and *Trioxys brevicornis* (Haliday). The mentioned parasites are all not strictly specialized and attack also other Myzine groups.

4. Tribe *Phorodontini*

Phorodon-species (*P. humuli* Schrk.) are dioecious, occurring in forest and intermediate habitats. On primary host plants and secondary host plants in these types of habitats they are attacked by *Ephedrus persicae* Frog., *Ephedrus plagiator* (Nees), which are widely specialized parasites occurring in forest and intermediate habitats. *Trioxys humuli* Mackauer, a third known parasite, is a strictly specialized species.

130

Myzodes-species are monoecious or dioecious, occurring in forest and steppe type of habitats. *Myzodes ligustri* (Mosl.) is a leaf-curling aphid. It is attacked in forest habitats by *Ephedrus persicae* Frog. and *Monoctonus cerasi* (Marshall), which are typical parasites of some leaf-curling aphid groups. Species of *Myzodes* in steppe habitats are attacked by the typical parasites – *Aphidius picipes* (Nees), *Diaeretiella rapae* (M'Int.) and *Aphidius matricariae* Haliday.

5. Tribe *Myzini*

Myzus-species *(M. cerasi)* are a dioecious species. In forest habitats they occur on primary host plants as leaf-curling aphids, being infested by a typical parasite complex of such aphids: *Ephedrus persicae* Frog., *Ephedrus plagiator* (Nees), *Lipolexis gracilis* Först., *Ephedrus cerasicola* Starý is apparently a specialized parasite.

6. Tribe *Cryptomyzini*

Capitophorus-species (*C. elaeagni* d. G.) are dioecious, occurring in forest and steppe habitats. In steppe type of habitats they are infested by a typical parasite of Myzine aphids-*Aphidius matricariae* Hal.

Myzella-species are dioecious, occurring in forest and intermediate habitats They are infested by a specialized parasite *Aphidius ribis* Hal.

Cryptomyzus-species, similar to *Myzella,* are dioecious and occur in the same type of habitats. The are attacked by the same specialized parasite.

7. Tribe *Nasonoviini*

Impatientinum-species are monoecious, occurring in forest type habitats (under-growth). They are attacked by a specialized parasite – *Monoctonus nervosus* (Hal.) that occurs in these habitats.

Nasonovia-species are monoecious or dioecious, occurring in forest and inter-mediate type of habitats. In forest habitats both the mono- and dioecious species are attacked by a rather specialized complex of parasites – *Monoctonus crepidis* (Hal.), *Aphidius hieraciorum* Starý, *Monoctonus angustivalvus* Starý, *Praon pubescens* Starý, which do not attack other aphid groups.

Hyperomyzus-species are monoecious or dioecious. The dioecious species (*H. lactucae* L.) in forest habitats are attacked by *Ephedrus plagiator* (Nees), which is a widely specialized parasite occurring in this type of habitats. On secondary host plants, in steppe habitats, they are attacked by *Aphidius sonchi* Marsh. which is a specialized parasite, and by *Praon volucre* (Hal.) and *Lysiphlebus fabarum* (Marsh.), which are widely specialized.

Rhopalosiphoninus-species are monoecious, living in forest habitats (undergrowth). They are attacked by *Ephedrus lacertosus* (Hal.), which is probably a specialized parasite.

5. Subfamily *Dactynotinae*

1. Tribe *Aulacorthini*

Microlophium-species (*M. evansi* Theob.) are monoecious, occurring in steppe type habitats. They are attacked by the same species as a number of other *Dactynotinae*, namely *Acyrthosiphon, Dactynotus* and *Macrosiphoniella* spp.

Acyrthosiphon-species are monoecious and they occur in forest or steppe habitats. In forest habitats living species *(A. spartii, A. caraganae)* are parasitized by typical widely specialized parasites occurring in forest type habitats: *Ephedrus plagiator* (Nees), *Trioxys angelicae* (Hal.), to lesser degree by a specialized parasite *Aphidius caraganae* Starý. In steppe habitats, occurring species are attacked by the typical steppe species – *Aphidius ervi* Haliday and *Praon dorsale* (Haliday).

Mirotarsus-species occur in steppe habitats and intermediate habitats and are attacked by a specialized parasite *Aphidius mirotarsi* Starý.

Aulacorthum-species are monoecious, occur mostly in forest and intermediate habitats, where they parasitized by probably a specialized parasite *Aphidius aulacorthi* Starý, to a lesser degree by *Ephedrus plagiator* (Nees).

Metopolophium-species are dioecious, occurring on primary host plants in forest type habitats and on secondary host plants in steppe type habitats. In steppe type habitats they are parasitized by *Aphidius avenae* Hal. and *Praon volucre* (Hal.), which attack some other *Dactynotinae* on grasses too.

2. Tribe *Macrosiphonini*

Linosiphon-species occur in forest habitats, being monoecious. They are attacked by *Aphidius matricariae* Hal., which is a parasite of Myzine aphids in steppe and intermediate habitats.

Macrosiphum-species are monoecious and dioecious, occurring in forest and steppe habitats. In forest habitats they are parasitized by more (*Aphidius rosae* Hal., *Praon rosaecola* Starý) or less specialized [*Aphidius rubi* Starý, *Aphidius lonicerae* Marsh., *Ephedrus lacertosus* (Hal.), *Ephedrus plagiator* (Nees)], parasite species. In intermediate habitats they are rarely attacked by steppe species *Aphidius ervi* (Haliday).

132

Sitobium-species are monoecious or dioecious, occurring in forest and steppe habitats. In forest type habitats they are parasitized on primary host plant by a typical widely specialized parasite *Ephedrus plagiator* (Nees). Monoecious species that occur in forest undergrowth are attacked by a typical parasite *Aphidius lonicerae* Marshall and the specialized parasite *Aphidius equiseticola* Starý and *Monoctonus caricis* (Haliday).

3. Tribe *Dactynotini*

Titanosiphon-species are typical monoecious inhabitants of forest type habitats. They are attacked by a typical parasite of Dactynotine aphids — *Aphidius absinthii* Marshall, *Praon absinthii* Bignell and by a strictly specialized parasite *Trioxys pannonicus* Starý.

Paczoskia-species (*P. major* Boern.) are monoecious, typical inhabitants of steppe habitats. They are attacked by a typical parasite species of Dactynotine aphids — *Aphidius funebris* Mack., *Praon dorsale* (Hal.).

Phalangomyzus-species [*P. oblongus* (Mordw.)] are monoecious, in steppe habitats occurring species. They are parasitized by a strictly specialized parasite *Aphidius phalangomyzi* Starý.

Macrosiphoniella-species are typical monoecious aphids, inhabitants of steppe type habitats. They are attacked by a typical complex of parasites: *Aphidius absinthii* Marsh., *Ephedrus campestris* Starý, *Praon absinthii* (Bignell), *Trioxys centaureae* (Hal.).

Dactynotus-species, with some exceptions, similarly as the species of the precedent genus, are typical monoecious aphids that occur in steppe habitats. They are parasitized by the typical parasite complex: *Aphidius funebris* Mack., *Ephedrus campestris* Starý, *Praon dorsale* (Hal.), *Trioxys centaureae* (Haliday).

Metopeurum-species are monoecious, in steppe type habitats occurring aphids. They are attacked by the strictly specialized parasite complex — *Lysiphlebus hirticornis* Mack., *Aphidius tanacetarius* Mack.

Microsiphum-species are also a typical steppe species, which are attacked by the same parasite as *Metopeurum*-species.

4. Tribe *Megourini*

Amphorophora-species occur in forest type habitats (undergrowth). They are attacked by a typical parasite *Aphidius lonicerae* Marshall, which attacks also other Dactynotine aphids in this type of habitat.

Nectarosiphum-species (*N. rubi* Kalt.) are monoecious, occur in forest type habitats (undergrowth, gardens, etc.). They are attacked by *Aphidius rubi* Starý, which parasitizes also some other aphids on *Rubus*.

Megoura-species (*M. viciae* Bckt.) are monoecious, occurring in forest meadows, pathways, etc. They are attacked by a specialized parasite *Aphidius megourae* Starý and by *Praon dorsale* (Haliday) that attacks other Dactynotine aphids too.

Conclusions:

Subfamily *Pterocommatinae* is a very natural group of aphids as to the general appearance and life-history. The parasites, being represented by one species — *Aphidius pterocommae* Ashm. are rather a specialized species that do not parasitize other aphid groups.

Subfamily *Aphidinae*. This is a natural relatively rather homogenous aphid group as to the general appearance and life history.

The tribe *Cryptosiphonini* has a peculiar position. Its members are attacked by parasites, which are apparently specialized more on the mode of host life as they attack some Myzine aphids too (*Ephedrus nacheri* Quilis).

As for other Aphidine aphids, a similar situation may be found in the genus *Schizaphis*, which is attacked by *Diaeretiella rapae* (M'Int.) that is the main parasite of some Myzine aphids, but also by *Ephedrus plagiator* (Nees), which is a typical parasite of *Aphidinae* and *Anuraphidinae*.

The greatest part of *Aphidinae* is infested by a parasite complex that includes parasites that attack also aphids of other groups, *Brachycaudina* and some *Myzinae*. The parasite complexes of the *Aphidinae* are strictly differentiated depending on habitats as it may be easily recognized from their composition on primary and secondary host plants. Strictly specialized complexes are comparatively rare, in such cases the given aphid species being usually attacked, besides a specialized parasite, by a widely specialized parasite too (e. g. *Aphis craccae*).

Subfamily *Anuraphidinae*. This subfamily, although it represents a natural aphid group, is not parasitized by a special complex of parasites. Its parasite attack also other aphid groups. A certain phylogenetic relation of the *Anuraphidinae* and their parasites is more or less distinct in forest habitats only. *Ephedrus persicae* Frog. is such a case, representing, however, a more widely specialized parasite. In steppe habitats occurring *Anuraphidinae* have no specialized parasites.

The parasites are distinctly differentiated as to the habitat so that the dioecious aphid species, which change the type of habitat (forest — steppe) because of migration, are attacked by different complexes of parasites that are typical for a given type of habitat.

Subfamily *Myzinae*. The Myzine aphids are attacked by a number of parasite species. The specialization of parasites is different. There are parasites, which are more or less specialized on a given Myzine group: On tribe — parasites of *Myzaphidini* or *Nasonoviini,* on group of genera — *Diaeretiella rapae* (M'Int.), *Aphidius matricariae* Hal., on genus — parasites of *Decorosiphon, Liosomaphis,* etc. to a lesser degree there are strictly specialized parasites. The influence of widely specialized parasites is apparent in the arboricolous aphid species in forest type habitats.

Subfamily *Dactynotinae.* It is parasitized by a number of typical species, the influence of widely specialized parasite species being less important.

The specific composition of parasites seems to be relatively more numerous in the tribes *Aulacorthini* and *Macrosiphonini* which include species that live in steppe and forest habitats. In forest habitats, besides specialized parasites, the influence of widely specialized parasites is apparent. In the steppe habitats the species of the mentioned aphid groups seem to have a more typical fauna of parasites.

The tribe *Dactynotini,* which includes typical species of steppe habitat fauna, has a corresponding typical steppe complex of aphidiid parasites. Species of this complex are able in some cases to parasitize also other aphid groups. The specific differentiation of parasites is very apparent. *Metopeurum* and *Microsiphum* are attacked by a rather typical and specialized parasite complex.

5. Family *Thelaxidae*

1. Subfamily *Anoeciinae*

Anoecia-species are not attacked by specialized parasite complexes. They are dioecious, for this reason there are different parasite complexes connected with habitats in which they occur and with modes of their life on primary and secondary host plants. *Anoecia*-species on primary host plants in forest type habitats live in dense colonies on tops of tree branches. There they are parasitized by *Lipolexis gracilis* Förster, which occurs mostly in steppe and intermediate habitats. On secondary host plants, in steppe habitats, they live on roots, being here parasitized by *Paralipsis enervis* (Nees), *Lysiphlebus dissolutus* (Nees), which are typical parasites of a number of root aphids.

2. Subfamily *Thelaxinae*

Glyphina-species are monoecious, occurring in forest type habitats (deciduous and mixed woods). There they are infested by *Aphidius sicarius* Mackauer.

Thelaxes-species occur similar to the monoecious species in forest type habitats (deciduous and mixed woods), being here parasitized by *Lysiphlebus thelaxis* Starý, which is a specialized parasite.

Conclusions: Each of the two subfamilies of the *Thelaxidae* is parasitized by different parasite complexes.

Due to the life-cycle, the *Anoeciinae* are attacked by different parasite complexes on primary and secondary host plants. Their parasites are not specialized, their host specificity being much wider.

The *Thelaxinae,* on the contrary, are a monoecious species. *Lysiphlebus thelaxis* Starý is a specialized parasite of *Thelaxes* spp., which has a certain taxonomic affinity to the parasites of the *Chaitophorus*-species. A part of the *Thelaxinae (Glyphina)* is attacked by widely specialized parasites for which the taxonomic affinity of the host seems to be less important (*Aphidius sicarius* Mack.).

6. Family *Eriosomatidae*

1. Subfamily *Eriosomatinae*

1. Tribe *Eriosomatini*

Schizoneura-species are dioecious aphids. On primary host plants they occur in forest type habitats (deciduous and mixed woods), producing galls on leaves of *Ulmus*. There they are infested by a specialized parasite, *Areopraon lepelleyi* (Waterston) and a widely specialized parasite *Ephedrus plagiator* (Nees). Both mentioned species are typical for forest type habitats.

2. Tribe *Byrsocryptini*

Byrsocrypta-species are dioecious aphids. On primary host plants in forest type habitats they produce galls on leaves of *Ulmus*. On secondary host plants, in steppe type habitats, they live on roots of grasses. On roots they are parasitized by *Paralipsis enervis* (Nees).

2. Subfamily *Pemphiginae*

1. Tribe *Pachypappini*

Asiphon-species occur in forest type habitats. On primary host plant — *Populus,* they produce galls on leaves; there they are attacked by a widely specialized *Ephedrus plagiator* (Nees), which is a typical parasite occurring in forest type habitats. On secondary host plants they occur on roots of coniferous trees; parasite complex is unknown.

Prociphilus-species are dioecious, occurring on primary host plants (*Fraxinus*) as gall-producing aphids in forest type habitats, where they are parasitized by *Ephedrus plagiator* (Nees). Their parasites on secondary host plants, on roots of conifers, are unknown in Czechoslovakia.

136

2. Tribe *Pemphigini*

Pemphigus-species produce galls on the leaves of primary host plants in forest type habitats. On secondary host plants they are poorly parasitized by *Lysiphlebus fabarum* (Marsh.) in steppe type habitats.

3. Subfamily *Fordinae*

In central Europe *Forda*-species are anholocyclic, occurring on roots of grasses in steppe habitats, where they are infested by *Paralipsis enervis* (Nees).

Conclusions: All of the above mentioned genera of the *Eriosomatidae* have a common feature in producing galls on primary host plants in forest type habitats. There they are parasitized by widely specialized parasites. The only exception is in the case of *Schizoneura,* which is attacked by a specialized parasite.

The dioecious species of the *Eriosomatidae* live on roots of secondary host plants. As far as it is known the species of *Byrsocrypta* and *Forda* are parasitized by *Paralipsis enervis* (Nees), which is a typical parasite of root aphids.

7. Family *Adelgidae*

Aphidiid parasites of this family are still unknown.

8. Family *Phylloxeridae*

Aphidiid parasites of this family are still unknown.

Ecological aphid groups and parasites

Aphids represent a rather variable group as to their ecology. The below mentioned ecological types of aphids are not strictly differentiable from each other as many intermediate species occur. Nevertheless, they are satisfactory for the separation of ecological groups of aphids and parasites.

The phylogenetical relationship of the host and parasite plays of course a certain role in different groups. For this reason the individual types are not subdivided depending on the importance of phylogenetical or ecological factors. In the greatest number of cases in parasites of the below mentioned aphids groups their ecological adaptation plays a major role, which results in that the parasites attack aphids of similar or the same mode of life but of different taxonomic affinity.

137

1. Gall producing aphids (Plate XVI, Figs. 17, 18).

Due to the sucking of aphids the host plant responds in producing a special formation — a gall. The gall is for a certain time entirely closed; this has a big importance for the adaptation of parasites as they can attack the gall-producing aphids only after the gall gradually opens and a few holes arise, through which the parasite female may penetrate into the gall.

There is a certain number of closely specialized parasites known (parasites of *Pemphigus* or *Forda*) but in Czechoslovakia only the less typical species have been bred till now.

Areopraon lepelleyi (Waterston) — *Schizoneura ulmi* (L.) on *Ulmus campestris*. This aphid species is a leaf-curling species, but the parasite is typical for the gall-producing *Schizoneura lanuginosa* (Hausm.) on *Ulmus*.

The gall producing aphids, whose galls are of a certain intermediary type between curled leaves and galls, are attacked by the same parasite as leaf-curling aphids (see 2) (E. g. the mentioned case of *Areopraon lepelleyi* Wat.).

2. Leaf-curling aphids (Plate XIII, XIV, Figs. 16, 20).

A number of aphid species from various groups falls into this group. The development of leaf-curling aphids is of a clearly convergent character. Due to the sucking of aphids the tops or leaves become curled and a certain kind of deformation is caused in this way. The curled leaf or top has then a certain protective significance for aphids against natural enemies. In the majority of cases the leaf-curling aphids are dioecious, causing curling of leaves of primary or secondary host plants.

The parasites are apparently rather adapted to such a mode of life as the taxonomic affinity to the host is clearly of secondary significance or not at all important in many cases.

As typical parasites of leaf-curling aphids the undermentioned parasites may be listed. Their more detailed ecological characteristics are presented in chapter IV. A part of the under mentioned parasites is, however, not typical (in parenthesis) as they infest previously free-living aphids. On the other hand, the typical leaf-curling parasites may also attack some free-living aphids, but they occur mainly as parasites of the leaf-curling aphid species.

Aphidius ribis Haliday — *Cryptomyzus ribis* (L.) on *Ribes rubrum*.

Areopraon lepelleyi (Waterson) — *Schizoneura ulmi* (L.) on *Ulmus*. (This is, however a typical parasite of gall-producing aphids.)

[*Diaeretiella rapae* (M'Int.)] — *Hayhurstia atriplicis* (L.) on *Chenopodium* spp.

Ephedrus cerasicola Starý — *Myzus cerasi* (F.) on *Prunus cerasus, P. avium*, etc.

Ephedrus nacheri Quilis M. P. — *Cryptosiphum artemisiae* (Bckt.) on *Artemisia vulgaris, Hayhurstia atriplicis* (L.) on *Chenopodium* spp.

Ephedrus persicae Frog. — *Rhopalosiphum padi* (L.) on *Padus racemosa, Aphis fabae* Scop. on *Euonymus europaea, Aphis idaei* (v. d. G.) on *Rubus idaeus, Roepkea marchali* (Börner) on *Prunus mahaleb, Allocotaphis quaestionis* (Börner) on *Malus silvestris, Dysaphis sorbi* (Kalt.) on *Sorbus aucuparia, Dysaphis devecta* (Walk.) on

138

Malus silvestris, Dysaphis spp. on *Malus silvestris, Pirus communis, Crataegus mono-gyna, Brachycaudus cardui* (L.) on *Prunus domestica, Prunus spinosa, Brachycaudus helichrysi* (Kalt.) on *Prunus spinosa, Brachycaudus* spp. on *Prunus domestica, Hyada-phis mellifera* Hottes on *Lonicera xylosteum, Myzodes ligustri* (Mosl.) on *Ligustrum vulgare, Myzus cerasi* (F.) on *Prunus avium.*

Ephedrus plagiator (Nees) — *Rhopalosiphum padi* (L.) on *Padus racemosa, Aphis fabae* Scop. on *Euonymus europaea, Aphis nasturtii* (Kalt.) on *Rhamnus cathartica, Aphis idaei* (v. d. G.) on *Rubus idaeus, Ceruraphis eriophori* (Walk.) on *Viburnum opulus, V. lantana, Dysaphis sorbi* (Walk.) on *Sorbus aucuparia, Dysaphis devecta* (Walk.) on *Malus silvestris, Dysaphis* spp. on *Malus silvestris, Pirus communis, Crataegus monogyna, Brachycaudus cardui* (L.) on *Prunus domestica, Prunus spinosa, Brachycaudus helichrysi* (Kalt.) on *Prunus spinosa, Prunus persica, Brachycaudus* spp. on *Prunus domestica, Myzus cerasi* (F.) on *Prunus avium, Hyperomyzus lactucae* (L.) on *Ribes nigrum, R. grossularia, Schizoneura ulmi* (L.) on *Ulmus, Prociphilus fraxini* (Htg.) on *Fraximus excelsior.*

[*Lipolexis gracilis* Förster] — *Myzus cerasi* (F.) on *Prunus avium, Prunus cerasus, Aphis fabae* Scop. on *Beta vulgaris, Brachycaudus helichrysi* (Kalt.) on *Prunus spinosa, Myzus cerasi* (F.) on *Prunus avium.*

Lysaphidus erysimi Starý — *Pseudobrevicoryne erysimi* Holman on *Erysimum crepidifolium.*

[*Lysiphlebus ambiguus* (Haliday)] — *Aphis schneideri* (Börner) on *Ribes rubrum.*

[*Lysiphlebus fabarum* (Marshall)] — *Aphis fabae* Scop. on *Beta vulgaris, Aphis rumicis* (L.) on *Rumex* spp.

Monoctonus cerasi (Marshall) — *Myzodes ligustri* (Mosl.) on *Ligustrum vulgare.*

[*Praon abjectum* (Haliday)] — *Aphis fabae* Scop. on *Euonymus europaea.*

[*Praon volucre* (Haliday)] — *Dysaphis* sp. on *Malus silvestris.*

[*Trioxys angelicae* (Haliday)] — *Aphis pomi* Deg. on *Malus silvestris, Dysaphis devecta* (Walk.) on *Malus silvestris, Dysaphis* sp. on *Malus silvestris, Pirus communis, Crataegus monogyna.*

Host and parasite list of leaf-curling aphids in Czechoslovakia.

Family *Aphididae*]

Subfamily *Aphidinae*

Rhopalosiphum padi (L.) — *Ephedrus plagiator* (Nees) on *Padus racemosa, Ephedrus persicae* Frog. on *Padus racemosa, Trioxys angelicae* (Haliday) on *Padus racemosa.*

Aphis fabae Scop. — *Ephedrus plagiator* (Nees), *Praon abjectum* (Haliday), *Trioxys angelicae* (Haliday) on *Euonymus europaea, Lysiphlebus fabarum* (Marshall), *Lipolexis gracilis* Förster on *Beta vulgaris.*

Aphis idaei (v. d. G.) — *Ephedrus persicae* Frog. on *Rubus idaeus, Ephedrus plagiator* (Nees) on *Rubus idaeus.*

Aphis nasturtii (Kalt.) — *Ephedrus plagiator* (Nees) on *Rhamnus cathartica.*

Aphis pomi (Deg.) — *Trioxys angelicae* (Haliday) on *Malus silvestris, Crataegus monogyna.*

Aphis rumicis (L.) — *Lysiphlebus fabarum* (Marshall) on *Rumex* spp.

Aphis schneideri (Börner) — *Lysiphlebus ambiguus* (Haliday) on *Ribes rubrum.*

Aphis viburni Scop. — *Trioxys angelicae* (Haliday) on *Viburnum opulus, V. lantana.*

Cryptosiphum artemisiae Bckt. — *Ephedrus nacheri* Quilis M. P. on *Artemisia vulgaris.*

Subfamily *Anuraphidinae*

Roepkea marchali (Börner) — *Ephedrus persicae* Frog. on *Prunus mahaleb.*

Ceruraphis eriophori (Walk.) — *Ephedrus plagiator* (Nees) on *Viburnum opulus, V. lantana.*

Allocotaphis quaestionis (Börner) — *Ephedrus persicae* Frog. on *Malus silvestris.*

Dysaphis devecta (Walk.) — *Ephedrus persicae* Frog., *Ephedrus plagiator* (Nees), *Trioxys angelicae* (Haliday) on *Malus silvestris.*

Dysaphis sorbi (Kalt.) — *Ephedrus persicae* Frog., *Ephedrus plagiator* (Nees) on *Sorbus aucuparia.*

Dysaphis spp. — *Ephedrus persicae* Frog., *Praon volucre* (Haliday), *Trioxys angelicae* (Haliday), *Ephedrus plagiator* (Nees) on *Malus silvestris, Pirus communis, Crataegus monogyna, Sorbus torminalis.*

Brachycaudus cardui (L.) — *Ephedrus plagiator* (Nees), *Ephedrus persicae* Frog. on *Prunus domestica, Prunus spinosa.*

Brachycaudus helichrysi (Kalt.) — *Ephedrus persicae* Frog., *Ephedrus plagiator* (Nees), *Praon volucre* (Haliday), *Lipolexis gracilis* Förster, *Lysiphlebus fabarum* (Marshall) on *Prunus spinosa, Prunus persica.*

Brachycaudus spp. — *Ephedrus persicae* Frog., *Ephedrus plagiator* (Nees), *Lipolexis gracilis* Förster on *Prunus domestica.*

Subfamily *Myzinae*

Hayhurstia atriplicis (L.) — *Ephedrus nacheri* Quilis, *Diaeretiella rapae* (M'Int.) on *Chenopodium* spp.

Pseudobrevicoryne erysimi Holman — *Lysaphidus erysimi* Starý on *Erysimum crepidifolium.*

Hyadaphis mellifera Hottes — *Ephedrus persicae* Frog. on *Lonicera xylosteum.*

Myzodes ligustri (Mosl.) — *Ephedrus persicae* Frog., *Monoctonus cerasi* (Marsh.) on *Ligustrum aviculare.*

140

Myzus cerasi (F.) – *Ephedrus persicae* Frog., *Ephedrus cerasicola* Starý, *Ephedrus plagiator* (Nees), *Lipolexis gracilis* Förster on *Prunus cerasus, Prunus avium.*

Cryptomyzus ribis (L.) – *Aphidius ribis* Haliday on *Ribes rubrum.*

Hyperomyzus lactucae (L.) – *Ephedrus plagiator* (Nees) on *Ribes nigrum, Ribes grossularia.*

Family *Eriosomatidae*

Subfamily *Eriosomatinae*

Schizoneura ulmi (Deg.) – *Areopraon lepelleyi* (Waterston) on *Ulmus* spp., *Ephedrus plagiator* (Nees) on *Ulmus* spp.

Prociphilus fraxini (Htg.) – *Ephedrus plagiator* (Nees) on *Fraxinus excelsior.*

3. In crevices of rind living aphids.

Aphids of this group live in deep crevices of rind which protect them from natural enemies because of the rather restricted space between the borders of a crevice that does not enable e. g. the oviposition as it is made by the greatest part of *Aphidiidae.*

Stomaphis-species. A close adaptation for the infestation of these aphids in the parasite *Protaphidius wissmannii* (Ratz.) had developed. It possesses a specifically morphologically adapted abdomen, the apical part of which is tubiform and prolongated and may be kept as a sham-ovipositor. Due to its length and movability it enables the infestation of aphids that are hidden in rind-crevices.

4. In leaf-sheaths living aphids

This group is characterized by having been partly protected against enemies by leaf-sheaths of the host plant *(Sipha, Thripsaphis, Hyalopteroides)* which leave only a small restricted space around the aphid.

This ecological aphid group does not seem to have specialized parasites.

Sipha-species are parasitized by *Lysiphlebus arvicola* Starý. Judging from the comparison with other groups of the *Chaitophoridae* – parasites it seems that the phylogenetical relationship is more important than the ecology in this case.

5. Root aphids (Plate XIII, Figs. 13, 14)

The root aphids may be divided into two groups:

A. Root collar aphids, which gradually spread to upper parts of the host plant.

B. True root aphids, which live on roots or root collars and do not spread to upper parts of the host plants.

The parasites infesting root aphids may be divided into two groups. In the members of these groups only a list of root aphids is given below, while the more detailed characteristics are given in chapter IV of this book.

a) To the first group belong the *Aphidiidae* of usual appearance that are not remarkably morphologically adapted to the existence in the nests of ants. They

141

infest mostly various species of free-living aphids, but they can also infest those inhabiting the runs of ants at the root collar. But the infestation of the root aphids seems to be of a lesser degree judging from their general host range. They have not been also ascertained as members of the ant-nest community.

The following species may be included in this group:

Lipolexis gracilis Förster — *Aphis taraxacicola* (Börner) on *Taraxacum officinale, Aphis* sp. on *Peucedanum alsaticum, Brachycaudus mordwilkoi* HRL. on *Echium vulgare.*

Lysiphlebus fabarum (Marshall) — *Aphis lambersi* (Börner) on *Daucus carota, Aphis neoreticulata* Theo. on *Fagopyrum convolvulus, Aphis plantaginis* (Goetze) on *Plantago major, Aphis poterii* Börner on *Sanguisorba minor, Aphis taraxacicola* (Börner) on *Taraxacum officinale, Aphis thomasi* (Börner) on *Scabiosa columbaria, Aphis vandergooti* (Börner) on *Achillea millefolium, Brachycaudus tragopogonis* (Kalt.) on *Tragopogon* sp., *Brachycaudus* sp. on *Senecio jacobaea, Protaphis carlinae* (Börner) on *Carlina acaulis.*

b) To the second group the aphidiids belong that are remarkably morphologically adapted and resemble in general the ants. The members of this group are typical parasites both of the true root aphids and on root-collar living aphids. They do not infest any free-living aphids on the upper part of plants. In some cases there is a very close relation of these parasites to aphid attending ants.

The following species fall into this group:

Aclitus obscuripennis Förster — This is a parasite of *Anoecia* and probably other species. It was, however, only swept in Czechoslovakia.

Lysiphlebus dissolutus (Nees) — *Protaphis* sp. on *Leontodon hispidum.*

Paralipsis enervis (Nees) — *Anoecia* sp. on *Agropyrum repens, Aphis roepkei* HRL. on *Potentilla anserina, Brachycaudus ballotae* (Pass.) on *Ballota nigra, Brachycaudus cardui* (L.) on *Carduus crispus, Brachycaudus mordwilkoi* HRL. on *Echium* sp., *Brachycaudus* sp. on *Arctium lappa, Chomaphis* sp. on *Arctium lappa, Dysaphis crataegi* (Kalt.) on *Daucus carota, Dysaphis subterranea* (Walk.) on *Pastinaca sativa.* (Note: A number of other root aphids is known from Europe, belonging to families *Aphididae, Thelaxidae, Eriosomatidae.*)

Host and parasite catalogue of root aphids in Czechoslovakia.

Family *Aphididae*

Subfamily *Aphidinae*

Aphis lambersi (Börner) — *Lysiphlebus fabarum* (Marshall) on *Daucus carota.*
Aphis neoreticulata Theo. — *Lysiphlebus fabarum* (Marsh.) on *Fagopyrum convolvulus.*

142

Aphis plantaginis Goetze — *Lysiphlebus fabarum* (Marsh.) on *Plantago media.*
Aphis poterii Börner — *Lysiphlebus fabarum* (Marsh.) on *Sanguisorba minor.*
Aphis roepkei HRL. — *Paralipsis enervis* (Nees) on *Potentilla argentea.*
Aphis taraxacicola (Börner) — *Lysiphlebus fabarum* (Marsh.) on *Taraxacum officinale, Lipolexis gracilis* Först. on *Taraxacum officinale.*
Aphis thomasi (Börner) — *Lysiphlebus fabarum* (Marsh.) on *Scabiosa columbaria.*
Aphis vandergooti (Börner) — *Lysiphlebus fabarum* (Marsh.) on *Achillea millefolium*
Aphis sp. — *Lipolexis gracilis* Först. on *Peucedanum alsaticum.*
Protaphis carlinae (Börner) — *Lysiphlebus fabarum* (Marsh.) on *Carlina acaulis.*
Protaphis sp. — *Lysiphlebus dissolutus* (Nees) on *Leontodon hispidum.*

Subfamily *Anuraphidinae*

Brachycaudus ballotae (Pass.) — *Paralipsis enervis* (Nees) on *Ballota nigra.*
Brachycaudus cardui (L.) — *Paralipsis enervis* (Nees) on *Carduus crispus.*
Brachycaudus mordwilkoi HRL. — *Lipolexis gracilis* Först. on *Echium vulgare, Paralipsis enervis* (Nees) on *Echium* sp.
Brachycaudus tragopogonis (Kalt.) — *Lysiphlebus fabarum* (Marsh.) on *Tragopogon* sp.
Brachycaudus sp. — *Lysiphlebus fabarum* (Marsh.) on *Senecio jacobaea.*
Brachycaudus sp. — *Paralipsis enervis* (Nees) on *Arctium lappa.*
Chomaphis sp. — *Paralipsis enervis* (Nees) on *Arctium lappa.*
Dysaphis crataegi (Walk.) — *Paralipsis enervis* (Nees) on *Daucus carota.*
Dysaphis subterranea (Walk.) — *Paralipsis enervis* (Nees) on *Pastinaca sativa.*

Family *Thelaxidae*

Subfamily *Anoeciinae*

Anoecia sp. — *Paralipsis enervis* (Nees) on *Agropyrum repens.*

6. Solitary aphids
No parasites of this group *(Hormaphidula)* have been bred in Czechoslovakia.

7. Aphids freely living in colonies
In this group the main part of aphids belong that have no remarkable adaptation. Their parasites are also rather numerous. Their specificity is influenced by a number of factors (see Ch. VII). Partly the complexes of parasites may be recognized from the relation of taxonomic aphid groups and their parasites.

8. Aphids that have saltatorial legs
Aphids of this group are rather mobile due to the peculiar saltatorial legs. This adaptation developed independently in various groups (*Drepanosiphum, Tinocallis, Izyphya, Saltusaphis,* etc.).

This aphid group does not seem to have a closely adapted parasite complex. Only a part of their parasites is closely adapted, while a number of them attacks also other aphids groups that do not possess saltatorial legs.

Drepanosiphum platanoidis (Schrk.). From its three known parasites only *Trioxys cirsii* (Curtis) seems to be more closely adapted, having developed, as all *Trioxys*-species, accessory prongs on the last abdominal sternite, which enables a more successful attack and oviposition. The other parasites, *Monoctonus pseudoplatani* (Marsh.) and *Dyscritulus planiceps* (Marsh.) do not possess any apparent morphological adaptation.

Tinocallis saltans (Nevs.). There is only one specialized parasite *Trioxys hortorum* Starý; the shape of its prongs seems to be an adaptation for enabling a more successful attack on the aphid. Other parasites, *Trioxys pallidus* (Haliday) and *Praon flavinode* (Hal.) attack also other *Callaphididae* that do not have saltatorial legs.

Phyllaphis fagi (L.). This aphid is attacked by a specialized *Trioxys phyllaphidis* Mackauer, whose accessory prongs have the usual purpose in enabling a more successful oviposition.

Izyphya, Saltusaphis, etc. They are attacked by specialized parasites that have, however, not been ascertained in Czechoslovakia till now.

Adaptations of parasites to aphids

A parasite adaptation to the host is a rather long-termed and complicated process in nature.

There is quite a number of parasite adaptations to the host, which may be classified in a different way in accordance with their degree, e. g. morphological or ecological adaptations. The rate of development of an adaptation may be longer or shorter, the adaptation being more or less apparent.

Morphological adaptations: Accessory prongs. In some genera accessory prongs on abdominal tergites or sternites develop. They enable a more successful attack and oviposition in the host that, being captured, cannot escape before a parasite egg is deposited. In the genus *Trioxys* two prongs on the last abdominal sternite are developed, by which the attacked aphid is held. In different species these prongs are of various lengths and pubescence, depending on host species or group on which the parasite is adapted. The parasite specialization is rather close in the majority of cases restricting on the one hand the parasite to a certain species or host group, on the other hand only a host aphid of certain instar may be attacked. As an example of the mentioned specialization or the close specialization of various species of the genus, *Trioxys* may be mentioned. A similar accessory prong, although being of a different shape as to the morphology, has convergently developed in the genus *Metaphidius*. In *Metaphidius aterrimus* (Fahr.) there is developed a tubiform prong on the base of tergite 7 which is used as a support against the upward curved ovipositor

144

sheaths. This is also an example of close parasite adaptation to the host with a similar result as in *Trioxys*.

Sham-ovipositor. *Protaphidius wissmannii* (Ratzeburg) is a parasite of *Stomaphis*-aphids that live in deep crevices of rind of various trees. The attack of such aphids would be difficult if the usual type of oviposition were used. For this reason probably the mentioned parasite had adapted to the host life by the development of a rather long and tubiform apical part of the abdomen that enables the attack of the host aphid at a relatively long distance. This part of the abdomen may be classified as sham-ovipositor.

Pubescence of ovipositor sheaths. Some parasites of wax-producing aphids have densely pubescent ovipositor sheaths [*Ephedrus validus* (Hal.), *Areopraon lepelleyi* (Waterston)]. This might be an adaptation in connection with wax covers of the host aphids but, on the other hand, in parasites of other wax-producing aphids a similar adaptation had not developed.

Mimetization. In case of mutualism of *Paralipsis enervis* (Nees), a parasite of underground aphids attended by *(Lasius)* ants, the similarity of parasite adults and ants is apparent. The similarity of heads both of ant and aphid parasite probably causes the parasite to be fed by ants as other symphils.

Physiological adaptations of parasites to host have been rather poorly known although their existence is certain and apparent in unnatural host propagation. Immunity reactions against egg and larvae of parasites may be mentioned, by the influence of which the parasite larvae are killed.

Behaviour. The host behaviour plays an important role in the adaptation of parasite. Such an adaptation includes on the one hand the adaptation to the general behaviour of the host, on the other hand the adaptation to the behaviour of different host instars, which is often different in accordance with host instar degree (escape reactions, movements, etc.).

In the case of parasite adaptation to the host's behaviour it seems that the adaptation to hosts mobility and escape reactions have the greatest significance. Quickly movable aphids are attacked by quickly moving parasites, the rate of attack and oviposition being shorter. Slowly moving aphids are attacked by slowly moving parasites, the rate of oviposition being longer in this case. The parasites are too adapted to the defensive movements of legs of attacked aphids. There is a behaviour adaptation also as to the responses to mechanical stimuli. This is apparent namely when behaviour of ant-attended and ant-non-attended aphid parasites is compared. Parasites of ant-attended aphids are often rather inactive when being influenced by mechanical stimuli, while parasites of ant-unattended aphids usually have quick escape reactions developed, the oviposition is broken off, etc.

A special example of behaviour, being a result of relation to ants, is the behaviour of *Paralipsis enervis* (Nees) adults. Antennal tapping was observed to be exchanged between *Paralipsis enervis* and *Lasius*-ant and the parasite was fed by regurgitation.

Ecological adaptations: Mode of host life. Typical ecological complexes of aphids — e. g. gall producing, leaf curling, root aphids, etc. are mostly attacked by specialized complexes of parasites.

Host life-cycle. Due to the life cycle of aphids, the parasites of dioecious aphids must still occur in the habitat after the emigration of aphids to another habitat. It seems they adapted themselves in two ways to such a host life-cycle. On the one hand their host-specificity is usually wider than in parasites of monoecious aphids, so that they are able to infest and develop on other aphids rather than on emigrated species. On the other hand, another response to the emigration of host aphids from the given type of habitat had developed in aphid parasites — the diapause. It is induced through the aphid host whose state is conditioned by the physiological state of the given plant due to the photoperiod, etc. This kind of adaptation enables the parasite to stay in a quiescent state during the period when a suitable host is lacking in the given type of habitat.

Host instar preference. The preference of lower aphid instar by the ovipositing parasite female seems to be also a kind of ecological adaptation due to the rate of host and parasite development as in many species the adult aphids (if attacked — in unnatural conditions) die before the parasite was able to finish its development. On the other hand, in some other parasite species higher instar or adult aphids are preferred due probably to the rate of host development. This may have a certain connection with the spread-possibilities of parasites.

Unnatural host propagation

Many aphid parasites are rather effective. Nevertheless, the economic value of some of these species is lowered by the fact that they are parasites of more or less economically indifferent aphids. For this reason there must be an effort in the research of the possibility of artificial change of the host specificity of such parasites through the unnatural host propagation. The results of such a successful effort, although the methods used are sometimes not easy, have to be the application of primarily economically indifferent parasites in the control of pest aphid species.

Before a certain aphid species is used in unnatural host propagation it is necessary to study the following:

1. Ecological characteristics of the parasite. This includes the knowledge of bionomics, distribution, habitat preference, host specificity, host × parasite relationship, etc. of the given parasite species.

2. Application of knowledge of host-parasite groups relationship. As studies have shown, there exist certain dependences that influence the infestation of certain aphid groups by parasites. If natural and unnatural host belong to two too different groups, the successful application is less probable or almost impossible. E. g. parasites of *Lachnidae* and *Aphididae*, etc.

146

3. Host morphology. The main morphological features of both hosts have to be at least similar. This depends also on the degree of parasite specialization.

4. Host behaviour. The behaviour of both natural and unnatural host has to be similar too. We can hardly succeed using quickly moving aphid species as unnatural host for a parasite of slowly moving aphids. Nevertheless, the intensity of aphid responses may be lowered artificially, using e. g. a lower temperature for a certain period, so that the aphid responds only slowly to the antennal tapping of the parasite and the aphid behaviour is a little changed.

5. Host size. As the aphid parasites prefer a species of a certain size, the size of unnatural and natural hosts has to be equal. Either the unnatural host aphid species or at least an instar of it have to be of a similar size to the natural host. In host size specialized species namely *(Trioxys)* this factor is rather important.

6. Stimulation effects of various factors on parasite female. Our observations have shown that it is advantageous to use for unnatural host propagation such parasite females that were kept isolated (fed by honey and water) for about 1 — 2 days without any contact with a host. It is recommended to remove from a Petri-dish aphid mummies, from which the parasites had emerged, too. In such a case the oviposition instincts are rather stimulated and the attack of unnatural host is more probable to be successful. In a certain case the female completely leaves the unnatural host. We found it to be advantageous to stimulate the female's oviposition instincts by transferring a specimen of a natural host aphid in the experimental Petri dish. The presence of the natural host causes strong stimulation of the oviposition instincts of the parasite female that begins to oviposit immediately in it. After several stings the natural host is removed. In sequence, many parasite females were then observed to oviposit in unnatural host too.

If it is possible because of the food specificity of the unnatural host aphid, it is advantageous to breed this host on the same host plant and in the neighbourhood of natural host aphid colonies. The odour and other factors of both aphids might become a little closer in this way at least in laboratory experiments.

7. Immunity reactions of the unnatural host. In some cases the mechanism of immunity reactions is important. The immunity reactions may be lower after several stings of the parasite female, after deposition of several eggs, and may be of different intensity in various instars, etc.

8. The successful laboratory propagation of an unnatural host aphid species does not, however, represent the real success in propagation as it is generally known that under laboratory conditions many hosts are attacked that would never be attacked in nature or in case of preference possibility. Therefore, only such laboratory unnatural propagation that continues to exist also in field experiments may be kept as quite successful. Further research as to the initial establishment etc. is then usual as in other aphid parasites.

VIII.

Superparasitism

In the case of superparasitism two or more primary parasite larvae of the same species are present in one host. It is caused by the oviposition habits of the female parasite that does not distinguish between parasitised and non-parasitised aphids. In consequence, one aphid may be infested several times by the same or other parasite females. This is common if a higher population density of parasites occurs.

In general, the following two cases of superparasitism may occur:

1. The eggs are laid in the same aphid almost simultaneously by one or more parasite females so that several larvae of the same age develop in consequence.

2. The eggs are laid in the same aphid by one or more females at certain intervals so that larvae of a different age develop.

The observations of many authors and ours too have shown that as only one larva completes its development, there is a competition among the larvae of the same or different instars in the same aphid. The mechanism of competition among the larvae is not equal; it depends on larval instar degree that compete. First instar larvae are remarkably mandibulate and there is a competition of a mechanical kind among them. The competition among the larvae of higher and lower instars is of physiological mechanism, as higher instar larvae overlive them exclusively, which have but negligible mandibles in comparison with first instar larvae. The physiological mechanism in the latter case is probably the lack of oxygen for the lower instar larvae, but might be of a food character, too.

Superparasitism may be commonly met with both in the laboratory breedings and in the field samples. In 1956, when studying the effectiveness of *Aphidius ervi* Hal. on *Acyrthosiphon pisum* in western Bohemia, we observed the common occurrence of 2–3 larvae of parasites in one aphid. In the laboratory breedings of *Aphidius megourae* Starý, a parasite of *Megoura viciae,* when dissecting aphids, we found the maximal number of parasite larvae in one host being 5 first instar larvae and one second instar larva. But through an experimental way a still higher number of parasite larvae in one host may be caused.

148

IX.

Aphid parasites and aphid attending ants relationship

The aphid × ant relationship may be generally of the three following types:

a) Constant. The relationship is continuous, resulting in adaptations both in morphology and ecology of the aphids.

b) Temporary. The aphids are attended by ants for some time in the season. Many holocyclic dioecious aphid species belong to this type. They occur e.g. in spring on leaves of plants, being attended by the ants there, in summer they migrate to stems and roots of other plants, where the ants do not attend them. In this case the relation between the ants and aphids is not very close.

c) Facultative. In this case the ants meet or occur in the colonies of aphid species belonging to this type quite occassionally and have no relation to them.

The aphids closely associated with ants have in general poorly developed structural modifications of a defensive character, if compared with those in the aphid species unattended by ants (cornicles, dense wax filaments, heavy sclerotization of cuticle, saltatorial legs, etc.). The constant aphid × ant relationship requires a certain type of the aphid's life-cycle. The obligatory host alternation, occurring in the holocyclic heteroecious species, prevents a continuous association with ants, while the holocyclic monoecious and anholocyclic species can associate continuously with the ants.

The division of parasites with regard to their relation to the aphid and ants relationship

The parasites can be divided into the following groups:

1. Parasites of aphids unattended by ants. It is quite a numerous group including many species whose hosts have no relation to ants, e.g. *Dyscritulus planiceps* (Marsh.), *Trioxys cirsii* (Curtis), *Monoctonus pseudoplatani* (Marsh.), *Aphidius funebris* Mackauer, *Aphidius absinthii* Marshall, *Praon dorsale* (Haliday), *Praon*

149

absinthii (Bignell), *Ephedrus campestris* Starý, *Trioxys centaureae* (Haliday), *Trioxys pannonicus* Starý, etc.

2. Parasites of aphids attended by ants. In this group there are two subgroups:

a) Non-specialized parasites. Their relation to ants depend more or less on their behaviour in relation to aphid hosts, and their relation to ants may be considered a response equal to other responses to the mechanical stimuli from the environment.

As mentioned above, the aphids attended by ants are rather poorly developed, as to their morpho- and ecological adaptation and behaviour. A similar behaviour has developed in the parasites, appearing in a long act of oviposition, negation of weaker mechanical stimuli, etc. *Lysiphlebus fabarum* (Marshall) can be presented as an example; its behaviour was observed in a colony of *Aphis fabae* on *Cirsium* sp., in July 1963 near Dresden (Germany). The aphid colony was numerous, with a high percentage of mummified aphids. It was densely occupied by ants and by about 8 adult specimens of *Lysiphlebus fabarum*. The parasites were newly emerged and intensively attacked the aphids of instar II. The ovipositing female parasites were often tapped by the ants' antennae, but they laid eggs without interruption (the oviposition lasted for about 40 seconds). The ants' behaviour was quiet, and the parasites were tapped in the same manner as the aphids. Being disturbed by our beating the plant with a pincette the parasites were quite indifferent and oviposited, while the ants responded in a completely different way, assuming a defensive attitude and shielding the aphids.

It results from our observation that the ants are indifferent to the aphid parasites, and shield them and the mummified aphids from strong adversary stimuli.

This character of the relation between the ants and aphid parasites is confirmed by a similar case observed in the same season and locality in the colony of *Chaitophorus* sp. on *Populus tremula*. The colony of *Chaitophorus* sp. was strongly parasitized by *Lysiphlebus salicaphis* (Fitch). The adult parasites, probably newly emerged, moved in the colony infesting the aphids. Two ant species — *Lasius* and *Formica* sp. — with a very different behaviour occurred in the colony. The ants of both species tapped the adult parasites with antennae, and they did not seem to be disturbed. When the plant was slightly beaten with the pincette, the ants reacted immediately and, especially *Formica*, assumed an aggressive attitude.

The important result of the observation is that the aphid parasitism does not depend on the presence or absence of ants, as the parasites disregard them. On the contrary the ants shield even the colonies of mummified aphids from enemies (hyperparasites were not considered). From this point of view the importance of the aphid parasites is even greater.

b) Specialized parasites. In the parasites of this group, besides the parasitism on aphids attended by ants, a particular relation to the ants has developed that can be considered as a case of mutualism. *Paralipsis enervis* (Nees) may be taken for a representative of this group. The ants behave to the adults of this parasite the

150

same as to the other symphils, palpating them; regurgitation was observed too. Besides, the ants mutilate the wings of parasites as it is usual in other cases of mutualism. The close relationship of *Paralipsis enervis* to the ants is apparent also in its morphological resemblance to the genus *Lasius*. Evidently it is a close adaptation to the most abundant aphid attending ant species, as *Paralipsis enervis* parasitizes quite a number of root aphids that are attended by *Lasius niger* particularly. Moreover, the movements of the adult parasites are typical too — the abdomen bent down in a specific way is quite unusual in the aphidiids, suggesting a complete adaptation of the parasite to the ant nest environment. *Paralipsis enervis* also hibernates in the adult stage in *Lasius* nests, while the other aphidiids overwinter in the praepupal stage inside mummified aphids. This too is an adaptation to the nest life and ant nest environment, as many species of underground aphids cannot survive in winter except with ants that are apparently adapted for tending them.

As the wings of the *Paralipsis enervis* female, searching for aphids in the ant nest, are mutilated, the parasite is unable to fly off and search for the aphids in another environment; thus its dispersal is secondarily restricted to the first-visited ant nest and depends on the dispersal of the myrmecophilous, particularly an underground aphid species dependent on ants in dissemination.

All morpho-ecological adaptations mentioned are well visible in the food-specificity of the parasite, as it attacks a number of root aphids attended by ants, if they live in underground ant chambers or inside ant runs around root collars of various plants.

Aphid attending ants and aphid parasites relationship

As our observations show, the ants ignore the presence of the adult aphid parasites. It results that the parasitization of aphids in the open does not depend on the presence or absence of the ants, being determined by other factors. That is why no adaptation has developed in the parasites even of a secondary character (mutalism in *Paralipsis enervis* × ants).

The ants protect the aphid colonies from the natural enemies, but ignore the existence of adult parasites in the colony. Thus they protect indirectly the parasites too from the natural enemies (aphid predators — see chapter X), safeguarding their ± undisturbed oviposition and thus increasing the percentage of parasitism.

The relation of the ants to the mummified aphids is similar. The ants attend them like the other aphids, leaving them untouched. An exception is *Lasius fuliginosus* with the aphid *Stomaphis* spp. mummified by the parasite *Protaphidius wissmannii* (Ratz.). In this case the mummified aphids are nibbled by the ants, so that only a shiny ball — a cocoon of the parasite remains. But the ants protect this cocoon in a similar way to the other ant species the usual mummified aphids. This unpublished observation of Prof. Dr Goidanich from the Torino University has been confirmed by the author's observation, and is apparent in the material of collected mummified aphids of *Stomaphis* spp., where the gradual nibbling of the aphid mummies by ants is well recognizable.

The ant and aphid and parasite association in relation to the natural limitation of aphids by parasites

Observations of various authors have shown that because of the ants' activity in increasing the environmental capacity for the attended aphids by removing honeydew or excavating galleries around suitable feeding sites the aphids can remain in favoured parts of the plant for a longer time and in a larger number than in the absence of ants.

The ant × aphid × parasite association may be in certain cases suitable for the natural limitation of aphids. As mentioned above, the aphid parasites are disregarded by the ants — it results that the aphids are parasitized whether the ants are present or not. In view of an increased degree of parasitism (factors affecting the food specificity of parasites, see chapter VII) the attendance of ants means a possibly prolonged occurrence of the aphids on a certain plant in a certain habitat, and thus a theoretical possibility of ants as attendants of the aphids on cultural plants, where the protection of the aphids from predators could have a negative effect, does not matter, the aphid × ant relation is actually advantageous in view of the natural limitation of aphids by parasites. Particularly if the aphids on the primary host plants are heteroecious, their relation to ants means a longer stay on the primary host plants. Regarding the fact that the aphids (in connection with temperature that they require for reproduction) usually have an advantage in number against the parasites, such a delay of the aphids is advantageous if the parasites occur, as it means a longer contact of the parasites with the aphids, i.e. increased parasitization and a decreased number of the aphids migrating to other habitats (particularly of the steppe type), often on cultural plants.

The close relation of the parasites to the ants attending the aphids, like that of *Paralipsis enervis* (Nees), is supposed to be less suitable for the dispersal and effectiveness of the parasite; since the wings of females are mutilated by ants, the dispersal of the parasite is secondarily limited to one ant nest, from where it can spread to the nearest neighbourhood only. But since *Paralipsis enervis* (Nees) was sometimes found to be a rather effective parasitic species, its life-history in relation to the influence mentioned and dispersal of newly emerged parasites from the ant nest in particular should be further studied, as it is the only parasitic species that can attack the underground aphids which often include pest species poorly attacked by other natural enemies.

152

X.

Natural enemies

The food chain of aphids and their natural enemies is rather numerous. It includes, besides the aphidiid parasites, quite a number of predators and parasites. The interrelationship among the natural enemies may be complicated in some cases. In this chapter the natural enemies are classified from the viewpoint of primary parasites — the *Aphidiidae*.

Plant kingdom

In the descendent stage of an aphid outbreak the aphid colonies are often infested by various fungi that either kill the aphids or develop on the honey-dew produced by aphids. Thus, aphid parasites i.e. their developmental stages that live inside aphids are killed commonly with them. There are some cases of a direct infestation of a *Praon*-cocoon by parasitic fungi known from literature, too. For the immediate influence on the aphid outbreak the activity of fungi may be classified as useful, but the problem of parasite control by fungi and its influence on further aphid outbreaks has not been solved up to now.

Animal kingdom

A. Parasites

Aa. Obligatory parasites.

a. Ectoparasites. Among rather common hyperparasites of aphids the chalcids of the family *Pteromalidae* belong, the species *Asaphes vulgaris* Walk., *Coruna clavata* Walk. and *Pachyneuron aphidis* Bouché being the most common. The bionomics of hyperpararasitic chalcids has many common features. The females attack parasitised aphids that include last instar larva or pupa of the aphidiid. The egg is laid on the surface of the primary parasite larva or pupa. The larva is cask-shaped, pointed at the apex. It is an external parasite that feeds gradually on the primary parasite larva and kills it and then pupates inside the aphid that had been primarily mummified by the aphidiid. The emergence hole has irregular margins, being easily recognizable and differentiable from the emergence hole of a primary parasite where the lid always occurs.

The encyrtid wasp, *Aphidencyrtus aphidivorus* (Mayr) is a rather common parasite of aphids, too. It is mostly mentioned as a primary or secondary aphid parasite in literature and it seems it may occur in both cases.

From the hymenopterous family *Ceraphronidae* the genus *Lygocerus* belongs among common parasites of the *Aphidiidae*. Females of these secondary parasites search for mummified aphids, which include the praepupa or pupa of a primary parasite. Their eggs are oval, slightly arcuate. They are laid on the surface of the host pupa or praepupa. The larvae live ectoparasitically and feed mostly on the thorax and abdomen of a primary parasite pupa. They are cask-shaped, somewhat narrowed at the apex. If more eggs were laid by the same or other females, they all develop but due to the competition only a single larva survives and reaches maturity. The mature larva pupates inside the primary parasite cocoon inside the mummified aphid. The imago cuts an emergence hole of irregular margins, similarly as in the chalcid hyperparasites.

In case that in the larva or pupa of an aphidiid (= primary parasite) there is a larva of a secondary parasite (entoparasitic charipine larva), they are both attacked and destroyed commonly by an ectoparasitic larva of a secondary parasite. Thus, both chalcid and ceraphronid larvae may sometimes occur as the tertiary parasites of aphids (see Fig. 7).

b. Entoparasites. From the cynipid subfamily *Charipinae* quite a number of species belong to the secondary parasites of aphids. The charipine female attack the living aphid that contains lower instar primary parasite larva, in which the egg is then deposited. The egg is oval, with a short process at one end. The instar I larva is rather sclerotized, with a long caudal prong on the last body segment. The mature larva is slightly arcuate. The larvae gradually feed on various organs and tissues of the primary parasite larva. By a certain mechanism the development of the host larva is somewhat prolongated but the larva does not pupate before being killed by the charipine larva. The mature charipine larva pupates inside the cocoon of the primary parasite — inside the mummified aphid. The emergence hole, similarly as in other secondary parasites, has irregular margins.

Ab. Facultative parasites.

a. Ectoparasites. Facultative ectoparasites of aphidiids are not common. Their facultative ectoparasitism has not been proved as their relationship to the aphidiid larvae in the same host is little known. Some species from the hymenopterous family *Aphelinidae* and from the dipterous family *Itonididae* belong to this group.

b. Entoparasites. Facultative entoparasites of the aphidiids are unknown.

154

B. Predators

Ba. Obligatory predators of the aphidiids are unknown.

Bb. Facultative predators.

In this group all the aphid predators may be included. The greatest part of them does not distinguish between parasitised and non-parasitised aphids, if the latter contains lower instar larvae of aphidiids, so that the parasite larvae are consumed commonly with aphids. The dead aphids, mummified by the parasites, are attacked by some predators only.

From the *Acari* there are some species of *Allothrombium* known as aphid predators. We observed them to be rather effective in the control of *Acyrthosiphon pisum* in a greenhouse and of *Megoura viciae* in the field.

From the class of insects quite a number of examples from various orders may be mentioned.

Neuroptera. Both the larvae and adults of the families *Chysopidae* and *Hemerobiidae* are sometimes rather effective aphid predators. According to our field observations, the larvae of *Chrysopa* attacked directly the mummified *Aphis fabae* aphids that contained last instar larvae or pupae of the aphidiid *Lysiphlebus fabarum* (Marshall) in a sugar beet field, where the aphids were heavily attacked both by parasites and predators. The long arcuate mandibles and the type of digestion enabled the chrysopid larvae to feed successfully on aphidiid larvae as they easily made small holes into the mummified aphids by their mandibles and then fed on parasite larvae. Nevertheless, it is believed the chrysopid larvae would prefer a living aphid in the case of the possibility of preference.

Thysanoptera. Some predatory thripids *(Haplothrips, Limothrips)* feed on aphids too.

Heteroptera. As the most common aphid predators of this order, nymphs and adults of the *Anthocoridae* and *Nabidae* may be mentioned.

Coleoptera. The most important group of aphid predators is represented by the larvae and adults of the family *Coccinellidae*. According to our observations both coccinellid larvae or adults are unable to feed on mummified aphids.

Hymenoptera. Some species of the family *Sphecidae* (genera *Nitela, Passaloecus, Diodontus, Stigmus, Pemphredon*) collect aphids from colonies and use them as food for their larvae in underground nests.

Diptera. Three main groups of predators may be recognized among the *Diptera* — the *Syrphidae, Itonididae* and *Chamaemyidae*. They are all predative in larval stage only, being mostly rather effective predators. According to our observations they are unable to feed on mummified aphids.

Besides the mentioned examples the class of insects includes many other aphid predators, which are however more facultative *(Orthoptera, Coleoptera — Staphylinidae, Carabidae)*.

As to the higher groups of animals, some birds *(Passeriformes)* and mice feed on aphids, too.

Effectiveness of natural enemies

From the mentioned groups of natural enemies the obligatory parasites *(Pteromalidae, Ceraphronidae, Cynipidae)* have the greatest importance. Their effectiveness is variable under various conditions. In some cases we observed almost 100% infestation of the aphidiids by secondary parasites. The percentage of infestation is especially great at the end of an aphid outbreak when primary parasites reached a higher level. In monoecious aphids, there is one outbreak throughout the season (Tab. 11), in dioecious aphids there is an outbreak on primary and then on secondary host plants.

Little is known about the field ecological characteristics of secondary parasites of aphids because of their poor taxonomic knowledge in many cases. Nevertheless, it seems there will be certain complexes too, but their specialization will be probably much wider than in the primary parasites — the aphidiids.

Table 11.

Effectiveness of secondary parasites on primary parasites of *Acyrthosiphon pisum*(Harris), western Bohemia, red clover fields, 1956

Species \ Date	6/7	19/7	1/8
Aphidius ervi	41 % 91·1	45 % 72·5	41 % 40·3
Praon dorsale	1 % 2·2	2 % 3·2	10 % 9·8
Alloxysta scutellata	2 % 4·4	10 % 16·1	15 % 14·7
Coruna clavata *Asaphes vulgaris*	1 % 2·2	5 % 8·0	33 % 32.4
Lygocerus sp.	— % —	— % —	3 % 2·9
Total number of mummified aphids	45	62	102

XI.

Foci in nature

The research of the foci of aphid parasites in nature is possible only on the basis of a rather deep study of aphid parasites in different types of habitats.

Considering that the aphid parasites of Europe have been carefully systematically studied during the past few years their main ecological features in field conditions are also known. For this reason, as the taxonomic research enabled further work, it was possible to approach a detailed analysis of parasite fauna of a different aphid species, the aphid pests namely, and their sources in nature. It is necessary to stress the necessity of complex research work without isolating aphid species from all the community, which is also so important for the research of mutual relations among various host × parasite species for biological control of aphids.

Terminology

A focus of an aphid parasite in nature is a part of the geographical landscape to which it is peculiar a certain biogeocoenosis, characterized by a more or less characteristic habitat and by the presence of biocoenosis, to which the host aphids belong.

Life-cycle of aphids and foci of aphid parasites

For the purpose of showing the relation between aphid life cycle and foci of parasites the division of aphids on monoecious and dioecious species is satisfactory.

Monoecious aphid species, in accordance with the peculiarities of their life cycle, do not change the type of habitat during the season as they do not migrate from primary to secondary host plants as the dioecious aphids do. With regard to the factors influencing the host specificity in the *Aphidiidae* (see Chapter VII)

157

monoecious aphid species are attacked by the same parasite complex during all the season, which are typical for the given type of habitat. The following examples may be mentioned here in this connection:

1. *Brevicoryne brassicae* L. is a monoecious aphid species occurring commonly on various *Brassicaceae* in steppe (field) type habitats. There it is infested by the parasite *Diaeretiella rapae* (M'Int.) which is a typical species for steppe type habitats.

2. Aphids of the genera *Dactynotus, Macrosiphoniella* are typical aphid species for steppe type habitats. There they are infested by a typical parasite complex that occurs in steppe habitats: *Aphidius funebris* Mackauer, *Aphidius absinthii* Marsh., *Praon absinthii* Bign., *Praon dorsale* (Hal.), *Trioxys centaureae* (Hal.), *Ephedrus campestris* Starý, etc.

Dioecious aphids, on the contrary, change the type of habitat during the season, as, in connection with the peculiarities of their life cycle, they migrate from primary to secondary and then to primary hostplants during the season. In different types of habitats they are infested by different parasite complexes, which are typical for the given type of habitat. The following examples may be mentioned:

1. *Anoecia* spp. occur in spring on primary host plants — *Cornus* in borders of woods, etc. where they are infested by *Lipolexis gracilis* Först. which is a typical parasite for steppe and intermediary types of habitats. At the beginning of summer they migrate to secondary host plants — on roots of grasses, where they are infested by *Paralipsis enervis* (Nees) which is a typical parasite species of root aphids in steppe type habitats.

2. *Sitobium granarium* (Kby) occurs in spring on primary host plants — on *Rosa* spp. in borders of woods, gardens, orchards, etc., in habitats of forest or intermediary type, where it is infested by *Ephedrus plagiator* (Nees), a typical parasite that occurs in forest type habitats. At the end of spring it migrates to steppe type habitats — on grasses, where it is attacked by *Aphidius avenae* Hal., which is a typical parasite occurring in steppe type habitats, and by *Ephedrus plagiator* (Nees) and *Praon volucre* (Hal.), which occur more rarely in steppe habitats, too, although they are a typical species for forest type habitats.

3. *Hyperomyzus lactucae* (L.) occurs in spring on *Ribes* spp. (as primary host plants) in forest type habitats such as orchards and gardens, parks, etc., where it is attacked by *Ephedrus plagiator* (Nees), which is a typical parasite that occurs in forest type habitats. In steppe type habitats, on secondary host plant — on *Sonchus oleraceus*, it is parasitised by *Aphidius sonchi* Marsh., which occurs as a typical species in steppe type habitats.

4. *Aphis fabae* Scop. occurs in spring on *Euonymus europaea* (primary host plant) in forest type habitats and intermediary habitats, where it is attacked by three parasite species — *Ephedrus plagiator* (Nees), *Trioxys angelicae* (Hal.), and *Praon abjectum* (Hal.), which are all typical species for forest type habitats. At the end of spring

the aphids migrate to secondary host plants — on sugar beet namely, and afterwards on various weeds, where they are infested by *Lysiphlebus fabarum* (Marsh.), a typical parasite species occurring in steppe type habitats, to a lesser degree by *Lipolexis gracilis* Först.

Types of foci

Various viewpoints may be used in the classification of foci of aphid parasites:

1. According to the number of parasite species the foci are divided into:

A. monospecific foci, if the aphid species is infested by a single parasite species only. Example: *Anoecia* sp. on roots of grasses in steppe habitats are attacked by a single parasite species — *Paralipsis enervis* (Nees).

B. bispecific or polyspecific foci, if the aphid is attacked by two or more parasite species in the given type of habitat. Example of bispecific focus: *Aphis fabae* Scop. is attacked in steppe type habitats by two parasite species — *Lysiphlebus fabarum* (Marsh.) and *Lipolexis gracilis* Först. Example — polyspecific focus: *Aphis fabae* is attacked in forest type habitats by three parasite species — *Ephedrus plagiator* (Nees), *Praon abjectum* (Hal.), *Trioxys angelicae* (Hal.).

2. According to the total rate of the focus existence:

A. Old foci are long-termed foci, a part of a more or less long-termed natural community with the natural balance. In Czechoslovakia, e.g. preservations of steppe landscapes, etc., where the fauna of both aphid and parasites is typical, as is clear from the comparison with big areas of steppes, e.g. in the USSR.

B. Recent foci are newly-formed foci, mostly short-termed, the origin of which is usually connected with activities of man. In this type of foci waste places, covered by a special plant association that is connected with a special aphid fauna, may be mentioned. It must be noted that just these types of habitats are usually inhabited in many cases by the fauna of old foci, which are preserved in natural conditions very poorly in Czechoslovakia, being usually more or less influenced by human activities.

3. According to the character of the origin of the focus:

A. Evolutionarily developed foci — autochthonous. In this type, foci developed during the evolution of long-termed community belong. Here again belong the more or less natural communities with natural balance such as woods, wood steppes, and steppes.

B. Anthropurgic foci had developed as a result of activities of man. Due to the activity of man, waste places, roadsides, etc., with characteristic fauna of aphids and parasites, have been found to be uncontrolled.

4. Intermediary foci. The different types of habitats cannot be sharply separated from each other but a special zone occurs between two different types of habitats, in which both the mentioned habitats have influence as to the fauna. In this intermediary zone intermediary foci of parasites occur, too. Example: *Aphis fabae* Scop. is attacked in the open field by two parasite species — *Lysiphlebus fabarum* (Marsh.) and *Lipolexis gracilis* Först. In case that a sugar-beet field is near a wood border, the aphids are attacked, at the borders of the field especially, by *Trioxys angelicae* (Hal.), which is however a typical parasite of *Aphis fabae* in forest type habitats. In this case, the influence of forest border on the parasite fauna of the sugar-beet field is clear.

5. Zoogeographical zones and the foci. The following types of landscapes may be recognized in the Palaearctic region, and with them the different types of foci are closely connected, too: a) tundra, b) taiga, c) forest steppe, d) steppe, e) semi-desert, f) desert, g) mountains.

Regarding the zoogeographical zones, the territory of Czechoslovakia is occupied by deciduous forest and groves zone, steppe and wood steppe zone, but various elements of tundra, taiga and semi-desert may be met with, too.

From the viewpoint of the classification used, the division of habitats on forest and steppe type habitats is satisfactory:

a) Forest type habitats. A forest represents a rather long termed community. The conditions in this community have relatively, therefore, a very poorly changeable character. The changes are mostly caused by the succession of various components of phytocoenosis. Indiscriminately, if the forests in Czechoslovakia represent more or less artificial monocultures or more or less natural forests, there is a common feature in forest type habitats, that within the frame of natural balance the parasites find their hosts every year in more or less the same habitat. In consequence, foci of parasites in this type of habitats are of a chronic type.

The elements of forest type habitats influence also in a different degree the fauna of parks and orchards, where the character of foci of parasites is in general the same as in forests.

As to the fauna of aphids and parasites, coniferous and deciduous forests are strictly differentiated. The latter have a special significance in the spread of the forest fauna, in cultural habitats of forest type (orchards). The analysis of foci is important in the classification of aphidofauna of orchards, i.e. sources of aphids and parasites, namely in the environment.

The forest fauna of aphids and parasites, with some exceptions, is not differentiated vertically as the same species may be found from mountains to lowland if the character of forest type habitats is preserved.

b) Steppe type habitats. In Czechoslovakia, like in the greatest part of European countries, the steppe zone is practically represented by the so called cultural steppe, while the rest of the true steppe may be found only in southern Moravia, southern

160

and eastern Slovakia and in some places in Bohemia. The general characteristic of steppe type habitats is as follows: They are primarily habitats of the steppe zone, habitats of the open landscape, mostly modified secondarily in the cultural landscape by the activity of man. For the purpose of this chapter the steppe type habitats may be further divided:

A. Steppe habitats that are under the direct and fairly important influence of human activity (fields). They are mostly monocultures that have a more or less limited composition of aphid fauna. In connection with the life-cycle of aphids this group may be subdivided:

Aa. Monocultures temporarily inhabited by a fauna of dioecious aphids. Example: *Sitobium* spp., *Metopolophium dirhodum*, *Sipha maydis*, *Rhopalosiphum padi*, etc. form the characteristic aphidofauna of a corn field. After corn, due to the crop rotation, peas, beans, etc. are grown. The aphidofauna of the latter mentioned plants is composed of *Aphis fabae*, *Acyrthosiphon pisum*, sometimes *Megoura viciae*, therefore, the composition of fauna is quite different from that of the corn field. This is of course recognizable also in the composition of parasites.

Therefore, it is apparent that the fauna of aphids and parasites changes in annual crop fields every year and is different from that of the preceding year. On annual crop fields the parasites a) do not have suitable conditions for overwintering due to agrotechnical activities (ploughing) of man, b) if they overwinter, they do not find a suitable host the following year as a new entomofauna of another cultural plant is formed with another specifically different aphid fauna composition, which is already not suitable for the given parasite species, c) the host migrates from the corn field at the end of summer, before winter, due to the ripe corn, and the parasites do not find another host here and they must spread to other places where a suitable host can still be found (in chronical foci of suitable habitats, etc.).

This type of habitat, therefore, does not practically include by its character any foci or it includes temporary foci only. On the contrary, the aphidiids move to these habitats in the course of the season — after the immigration of the host — from the neighbouring habitats that include chronic foci of parasites.

Ab. Monocultures inhabited by a monoecious aphid fauna. As a classical example of this type, perennial fodder crops (*Medicago*, etc.) infested by *Acyrthosiphon pisum* and *Therioaphis* may be mentioned. The aphids occur in the field during the whole season. The parasites may concentrate and hibernate here, and in the following year they may find again the same host species in the same habitat. Agrotechnical activities (cutting) destroy a part of the developmental stages of the aphidiids, but these activities are never so deep and destructive as ploughing which has a deep and destroying influence on the whole community. Under the influence of community-destroying activities the aphidiids are, in the case of perennial fodder crops, affected after two years in this habitat. Therefore, habitats of this kind include chronic foci of parasites.

B. Steppe habitats that are not exposed to the direct and effective influence of human activities (pastures, virgin lands, waste land, roadside, etc.). These habitats are often inhabited by a rather heterogeneous plant community, with which a fairly heterogeneous aphid fauna is connected. The habitats include chronic foci of parasites.

6. According to the rate of the existence of the focus within one year:

A. Temporary foci. They occur temporarily in the course of a year. Example: In corn fields groups of *Cirsium* plants as weeds infested by *Aphis fabae* Scop. are common. The aphid is infested by *Lysiphlebus fabarum* (Marsh.), which at first immigrates here, similarly as aphids, and moves from here after the emigration of aphids.

B. Chronic foci. They represent sources of parasites during the whole period of their existence. Example: A potato field is attacked by *Myzodes persicae* Sulz., or, a rape field is attacked by *Brevicoryne brassicae* L. The neighbourhood of these fields waste places or roadsides, etc. are covered with various weed plants, e.g. *Atriplex, Chenopodium*, etc. which are infested by *Hayhurstia atriplicis*. In relation to *Myzodes persicae* such places represent monospecific foci of a chronic type as *Hayhurstia atriplicis* is infested by *Diaeretiella rapae* (M'Int.), which spreads from here to rape − and potato fields and attacks *Myzodes persicae* Sulz.

7. According to the economic significance of the specific composition of parasites:

A. Indifferent foci contain parasites of economically indifferent aphid species, which do not include any economically significant aphid species within the range of their food specificity. Example: Waste places are sometimes covered with *Achillea, Artemisia* or *Tanacetum*. The aphidofauna of these plants and its aphidiid parasites are rather specialized and they do not have, being entirely economically indifferent, any relation to the activity of man.

B. Useful foci include parasites of pest aphids. Example: Field boundaries are often covered with *Salvia* spp. that are infested by *Aphis salviae, Plantago* sp. infested by *Aphis plantaginis, Tragopogon pratense* − infested by *Brachycaudus tragopogonis;* all the mentioned aphids represent alternative hosts of *Lysiphlebus fabarum* (Marsh.), and of *Lipolexis gracilis* Först., which also parasitises quite a number of pest aphids on cultural plants in the environment − e.g. *Aphis fabae* Scop., *Aphis craccivora* Koch, *Brachycaudus cardui* L., etc.

Moreover, the above mentioned useful aphid species occur mostly in spring on root-collars and enable the parasites to reach a certain population level before pest aphids like *Aphis fabae* (dioecious) migrate to field habitats.

C. Noxious foci are very rare. Some *Cinara* spp. occurring in woods could be mentioned, as these aphids are producers of honey dew. But the quality of forest-honey is often discussed so that the problem has not yet been clarified.

8. The end of foci of aphid parasites

The foci cease to exist in the following cases:

A. The hosts (all the hosts in the case of widely specialized parasites) or parasites fall out of the community due e.g. to chemical treatment.

B. The territory of the focus changes (drying, irrigation, burning, etc.), in consequence of which also another fauna occurs.

C. The fertilization of fallows (ploughing), particularly in chronic foci.

D. The factors of microclimate and particularly of humidity sharply change (partial moving).

A review of the aphidiid fauna of different types of habitats with respect to the classification of foci

It is difficult to separate various types of habitats into several groups in a Central European country as the individual types of habitats represent comparatively small areas in comparison with big complexes of communities as can be found e.g. in Central Asia.

For this reason we used the division of habitats into two main groups i.e. forest (1 – 12) and steppe (13 – 22) type habitats, which are subdivided into individual kinds of habitats.

In individual types of habitats the main and typical parasite species in connection with their aphid hosts are mentioned. Furthermore, the main types of foci, relation to the environment, etc. are classified.

There is no doubt in many habitats identical or similar aphidiid species occur depending on their ecological requirements. Nevertheless, the division of habitats used is supposed to be advantageous as the requirements of individual species of the *Aphidiidae* are various and different, so that the application of such a viewpoint would be hardly valuable.

1. Spruce forest

The spruce forests are typical mostly for mountain and submountain districts in Czechoslovakia. They represent monocultures for the main part, to a lesser degree they occur in mixed forests.

Various *Cinara*-species that attack spruce occur mostly in young trees in clearings, etc. There they are also attacked by parasites of the genus *Pauesia* [*P. piceaecollis* (Starý), *P. pinicollis* (Starý)], which are typical inhabitants of these habitats.

Pauesia-species have a certain relation to the aphidofauna of other coniferous forests as they are known also as parasites of other *Cinara*-species which occur there. *Pauesia*-species are specialized parasites that do not parasitize other aphid groups.

As a spruce forest represents a perennial community, foci of a chronical type occur here.

The economic importance of *Pauesia*-species in the mentioned type of habitats is small because of the small significance of their hosts too.

2. Fir forest

Fir forest complexes are comparatively rare today in Czechoslovakia. They occur for the main part in mountain and submountain districts.

Some species of *Pauesia,* parasites of Lachnid aphids infesting fir trees, are typical for fir forests. They are: *Pauesia grossa* (Fahr.), a specialized parasite of *Todolachnus abieticola* (Chol.), *Pauesia infulata* (Haliday), a parasite of *Buchneria pectinatae* (Nördl.).

Pauesia-species infesting fir aphids have a [*P. infulata* (Hal.)] certain relation to the parasites of other coniferous trees infesting aphids in accordance with their specificity. Otherwise their relation to the aphidofauna of a non-coniferous forest is negligible as they represent a rather specialized group.

As in other coniferous forests, in the fir forest the foci of parasites are of a chronic type also.

The above mentioned *Pauesia*-species were sometimes observed as effective parasites that heavily attacked colonies of lachnids.

3. Larch forest

Larch forests are typical for mountain and submountain districts. Larches may often be found in complexes of other trees too.

The typical aphidiid parasite complex that occur in these forests is formed by *Pauesia*-species, parasites of *Cinara*-species infesting larch.

Because of the host specificity of these *Pauesia*-species there exists a certain connection with other types of coniferous forests, as some *Pauesia*-species [*P. pini* (Haliday), *P. abietis* (Haliday), *P. laricis* (Haliday)], attack also the pine infesting *Cinara*-species.

The foci of parasites in a larch forest, similarly as in other coniferous forests, are of a chronical type.

Pauesia-species parasitizing *Cinara*-species that occur on larch were often observed to be highly effective as they destroyed almost entirely colonies of aphids occurring on larch.

4. Pine forest

Pine forests are untypical as to their occurrence as they occur in rather various landscapes.

Various pine species may be found from mountains to lowland, from wet to dry (sandy) habitats. Similarly, both their aphidofauna and its parasites occur, with some exceptions, independently on the above-sea-level altitude. The same parasite species of pine-infesting aphids may be found everywhere in a pine forest [*Pauesia pini* (Hal.), *P. abietis* (Hal.), *Metaphidius aterrimus* (Fahr.), *P. silvestris* (Starý), *Diaeretus leucopterus* (Hal.), *Praon bicolor* Mack., *Pauesia unilachni* (Gahan)].

A part of the parasite complex that is connected with a pine forest has connections with other coniferous forest communities, where a group of species [*P. pini* (Hal.), *P. abietis* (Hal.), *P. laricis* (Hal.)] infest certain *Cinara*-species. Another part, which is represented by specialized parasites of some on pine occurring aphids *(Protolachnus, Schizolachnus)*, i.e. parasite species *Diaeretus leucopterus* (Hal.), *Praon bicolor* Mack. and *Pauesia unilachni* (Gahan) are typical only for a pine forest community.

The forest type habitat, to which also a pine forest belongs, determines the chronic character of parasite foci.

Parasite species of the pine forest complex, both parasites of *Cinara*-species or *Protolachnus*- and *Schizolachnus*-species were observed occasionally to reach high effectiveness.

5. Juniper shrubs

Juniperus shrubs may be commonly met with in various submountain and mountain pasture meadows.

Their aphidofauna is formed by the *Cupressobium*-species, which are attacked by 2 specialized parasites in Czechoslovakia: *Pauesia juniperorum* (Starý) and *P. cupressobii* (Starý).

Because of the specificity of the mentioned parasite-species there is no relation of *Cupressobium* parasitizing *Pauesia*-species to other coniferous forest communities. Nevertheless, data from other countries show that in some cases *Pauesia*-species occurring in other coniferous communities [*P. infulata* (Hal.)] may attack also *Cupressobium*-species on *Juniperus*.

The foci of parasites in *Juniperus* shrubs are of a chronical type.

The economic importance of *Cupressobium* parasitizing *Pauesia*-species, similar to that of their hosts, is negligible.

6. Mountain and submountain forest and pasture meadows
(Plate XVI, Fig. 27)

In this type of habitat there usually occur, especially in boundaries, elements of parasite-complexes of forest, meadow and waste place, etc. habitats. The specific composition of parasites depends on the environmental habitats, e.g. species *Ephedrus plagiator* (Nees), *E. persicae* Froggatt, *Aphidius avenae* Hal., *Monoctonus nervosus* (Hal.), *Pauesia*-species, etc. can be found in these habitats.

7. Deciduous forest (Plate XVI, Fig. 28)

The deciduous forest covers the greatest part of Czechoslovakian forest districts. In the greatest part of these districts in Bohemia and Moravia there are mixed forests, while the Carpathian mountains in Moravia and Slovakia (except the High Tatra and Fatra mountains, etc.) are covered with beech forest. The deciduous woods form a number of intermediate habitats such as shrubs in fields, along pathways and ditches, etc. from where the elements of corresponding parasite complexes spread in the neighbourhood.

The parasite complexes that occur in deciduous forests are rather numerous. In a number of cases they are strictly specialized or, more often, the specificity of their representatives is relatively wide.

A typical parasite complex form parasites of aphids that occur on *Acer*-species, which is a rather common tree in deciduous woods. There are either parasites of *Drepanosiphum*-aphids [*Dyscritulus planiceps* (Marsh.), *Monoctonus pseudoplatani* (Marsh.), *Trioxys cirsii* (Curtis)], or of *Periphyllus*-aphids (*Trioxys falcatus* Mack., *Aphidius setiger* Mack.). They all occur commonly in deciduous woods, field shrubs, gardens, etc. and are widely distributed in similar habitats such as parks, shady tree avenues, etc.

Parasites of aphids associated with *Salix*-species represent also a typical deciduous forest complex. *Aphidius pterocommae* Ashm. is a parasite of *Pterocomma*-species, *Lysiphlebus ambiguus* (Hal.) is a parasite of *Aphis farinosa* (Gmel.). Both parasites mentioned are very common in various types of deciduous forest and intermediate type habitats where *Salix*-trees occur.

Parasites associated with *Populus*-aphids *(Chaitophorus)* — *Lysiphlebus salicaphis* (Fitch) are, similar to members of the above mentioned complex, typical members of deciduous forest habitats.

A number of aphids associated with *Betula* are attacked by *Aphidius sicarius* Mack. *(Betulaphis, Callistaphis)*. Another parasite, *Trioxys betulae* (Marsh.), is a specialized parasite of *Symydobius betulae*.

Lysiphlebus thelaxis Starý, a typical inhabitant of deciduous forests, a parasite of *Thelaxes*-species, is a specialized parasite. Other aphids infesting *Quercus* are attacked by other, more widely specialized parasites.

166

Aphidius hortensis Marsh., a parasite of *Liosomaphis berberidis, Aphidius ribis* Hal., a parasite of *Cryptomyzus ribis* on *Ribes* spp., *Aphidius rubi* Starý, a parasite of *Rubus* aphids such as *Macrosiphum funestum* and *Nectarosiphum rubi,* are typical parasites occurring in shrub undergrowth and borders and clearings of deciduous forests.

Protaphidius wissmannii (Ratz.) is a typical deciduous forest inhabitant, being known as a parasite of the *Stomaphis*-species, which occur on various trees- *Salix, Quercus, Betula, Fraxinus*, etc.

The beech forests are very poorly inhabited both as to the aphid species and their parasites. The aphid *Phyllaphis fagi* (L.), a common pest of the beech tree in Czechoslovakia, is attacked by *Trioxys phyllaphidis* Mack., a strictly specialized parasite species.

A number of Callaphidid species, occurring on *Quercus, Tilia, Juglans, Ulmus*, etc. is attacked by two typical inhabitants of deciduous forest habitats, by *Trioxys pallidus* (Hal.) and *Praon flavinode* (Hal.).

Areopraon lepelleyi (Wat.) is a specialized parasite of gall aphids and leaf-curling aphids *(Schizoneura)*.

Ephedrus persicae Froggatt and *Ephedrus plagiator* (Nees) are common parasites of the leaf-curling aphid species namely (*Dysaphis, Brachycaudus, Myzus, Schizoneura, Prociphilus*, etc.), which occur on various deciduous trees and shrubs (*Sorbus, Malus, Pirus, Prunus, Euonymus*, etc.).

Praon abjectum (Hal.), *Trioxys angelicae* (Hal.) and *Praon volucre* (Hal.) are also characteristic members of one parasite complex occurring in deciduous forests. *Praon abjectum* (Hal.) is a parasite of many *Aphis*-species, *Trioxys angelicae* (Hal.) of *Aphis, Brachycaudus*, etc., *Praon volucre* (Hal.) is a widely specialized parasite.

In deciduous forests, especially in their boundaries, some elements of steppe parasite complexes, like *Lipolexis gracilis* Först., sometimes occur, which attack e.g. *Myzus cerasi* (F.) on *Prunus avium*.

The greatest part of the parasite complexes mentioned is typical for deciduous forest habitats. They may be met with mostly everywhere in deciduous forests. On the other hand, a lesser part of parasites, *Trioxys angelicae* (Hal.), *Praon abjectum* (Hal.) are more connected with boundaries of forests, sunny forests and various shrubs that usually occur in intermediate habitats. This is, in connection with their specificity, an occurrence of aphid host plants, etc. They are also typical for groups of shrubs in fields. From these habitats they can penetrate to environmental steppe habitats. Some species, the ecological plasticity of which is wider, like *Ephedrus plagiator* (Nees), occur in forest undergrowth, deciduous forest habitats, intermediary habitats, and to a certain degree also in steppe habitats, where suitable hosts may be found too.

The foci of parasites in deciduous forests are of a chronical type as in other forest type communities.

The economic importance of individual species is various. In many cases the above mentioned parasite species are rather effective but the economic importance of their hosts is not so great. In certain cases, when an aphid outbreak causes serious damage in deciduous forests e.g. *Phyllaphis fagi* (L.) on beech, the parasites might have a certain significance if a more detailed research is made.

In every case, parasites of dioecious aphids which occur in deciduous forests (boundaries) on primary host plants are important as they kill a number of aphids which then migrate to cultivated plants in steppe type habitats [*Ephedrus persicae* Froggatt, *Ephedrus plagiator* (Nees), *Trioxys angelicae* (Hal.), *Praon abjectum* (Hal.), etc.].

8. Forest undergrowth (Plate XVI, Fig. 28)

The forest undergrowth of fir, spruce and deciduous woods has many common features both as to the aphids and parasites. On the contrary, the undergrowth of pine forests, although there are some similar features too, has the composition of aphidofauna and its parasites highly influenced by the environmental habitats.

A rather typical parasite-complex that occurs in forest undergrowth are *Nasonovia*-species parasites: *Aphidius hieraciorum* Starý, *Praon pubescens* Starý, *Monoctonus crepidis* (Hal.), *Monoctonus angustivalvus* Starý. They penetrate also in parks and gardens.

Another two typical parasite species are *Aphidius lonicerae* Marsh. and *Ephedrus plagiator* (Nees). *Aphidius lonicerae* Marsh. attacks many aphids on various plants in undergrowth: *Amphorophora ampullata* Bckt. and *Aulacorthum dryopteridis* Holman on *Dryopteris austriaca, Macrosiphum gei* (Koch) on *Geum, Macrosiphum prenanthidis* Börner on *Prenanthes purpurea, Macrosiphum daphnidis* Börner on *Daphne mezereum.* Although the second mentioned species, *Ephedrus plagiator* (Nees), attacks many aphids living on trees, it is also a parasite of *Macrosiphum prenanthidis* Börner on *Prenanthes purpurea.*

Aphidius equiseticola Starý, a parasite of *Sitobium equiseti* Holman on *Equisetum silvaticum,* is a specialized parasite.

Monoctonus nervosus (Haliday), a parasite of *Impatientinum balsamines* (Kalt.), may be found in undergrowth of *Impatiens*-species.

In undergrowth *Ephedrus lacertosus* (Hal.), a parasite of *Rhopalosiphoninus* sp. on *Oxalis,* also occurs.

Because of the general type of the forest community also the forest undergrowth includes chronical foci of parasites.

The greatest part of parasite complexes that occur in forest undergrowth has no economic importance. Only parasites of the *Nasonovia*-species could serve as a source for collecting material for biological control of *Nasonovia ribisnigri,* which occurs as a pest of some plants in gardens.

168

9. Peat bogs (Plate XVIII, Fig. 34)

Peat-bogs may be found both in the mountains and lowlands in Czechoslovakia.

Parasite complexes occurring here are rather typical, being formed by members of the genus *Diaeretellus* Starý [*D. ephippium* (Hal.), *D. heinzei* (Mack.), *D. macrocarpus* (Mack.)]. They are all parasites of aphids living on *Sphagnum* and various mosses, to a lesser degree also of others on peat-bog living aphid species like *Rhopalosiphum nymphaeae* (L.) on *Alysma plantago*, etc.

In old lowland peat-bogs *Typha angustifolia* often occurs, being infested by *Schizaphis scirpi* (Kittel). The aphid is parasitized here by *Ephedrus plagiator* (Nees), a parasite of a number of various aphid species in forest habitats.

The foci of *Diaeretellus*-species on peat-bogs are of a chronic type because of the perennial character of the community.

10. Orchards, orchard-avenues, vineyards (Plate XVI, Fig. 29)

They are habitats of a more or less deciduous forest type, but the aphidofauna is somewhat different due to the growing of many trees that do not occur in natural communities (*Prunus armeniaca, Prunus persica*, etc.) in Czechoslovakia. There are either cultivated orchards, or avenues of fruit trees, or groups of trees are grown in vineyards in a rather typical kind of southern Moravia and Slovakia habitats.

The parasite complex is to a certain extent similar to that of the deciduous forest. It consists mostly of *Ephedrus persicae* Froggatt, a parasite of *Dysaphis*-species on *Malus, Pirus*, and *Myzus cerasi* on *Prunus avium* and *Prunus domestica*, *Brachycaudus*-species on *Prunus*-spp., *Phorodon humuli* Schrk. on *Prunus domestica*, etc., *Ephedrus plagiator* (Nees), which has a similar specificity. The latter species, besides, attacks *Hyalopterus pruni* (Geoffr.) on *Prunus domestica*. *Ephedrus cerasicola* Starý is probably a specialized parasite of *Myzus cerasi* (F.) on *Prunus*-species. *Praon volucre* (Hal.) is another parasite of *Hyalopterus pruni* (Geoffr.) in orchards. *Trioxys angelicae* (Hal.) occurs sometimes here as a parasite of the *Dysaphis*-species and often as a parasite of *Aphis pomi* Deg. on *Malus silvestris*. *Lipolexis gracilis* Förster is an occasional parasite of *Myzus cerasi* (F.) on *Prunus avium* in orchards.

As orchards and avenues represent perennial communities, the foci has to be of a chronical type. Nevertheless, on the one hand there are parasites of dioecious aphids that occur here for a certain part of the season only. For this reason the composition of the orchards' neighbouring plants and other aphidofauna is important as there other suitable hosts of parasites of dioecious orchard aphid pests may be found. On the other hand, the specific features of some parasite species (*Ephedrus persicae* Frog.) show that such hosts are not necessary as the parasites fall in diapause after emigration of aphids.

The economic importance of a number of aphid parasites is great as they are parasites of serious aphid pests that may be often controlled by chemical treatment

169

sometimes with difficulties, being a leaf-curling species. Nevertheless, in certain cases the ecological adaptation of parasite species (diapause) reduce their effectiveness because of the high percentage of occurrence of diapausing cocoons at the maximum point of the aphid pest outbreak.

11. Parks, shady trees avenues, ornamental gardens

The above mentioned habitats are habitats of the forest type, the composition of whose plant community is mainly artificial. Nevertheless, a number of plant species occur commonly in nature, both in coniferous and deciduous forests. Besides, a part of trees and shrubs that are grown in parks, cannot be met with in nature (*Philadelphus coronarius,* etc.). Similarly, the aphidofauna is for the main part identical with that of forest type habitats (see previous chapter). The parasites occurring in parks can be commonly met with in deciduous forests too, but in parks they attack a number of aphid species that occur more rarely in nature.

In general the above mentioned habitats are characterized by rather numerous aphid fauna due to the numerous species of trees and shrubs that are grown in one place.

Similarly as in deciduous woods also the parks include chronical foci of parasites. The foci are of economic value or indifferent.

The economic importance of parasites occurring in parks is of the same degree as in deciduous forests. In certain cases their effectiveness is valuable in the case of park and ornamental tree pest aphids.

12. Hop gardens

Hops are grown mostly in Northwestern Bohemian districts, to a lesser degree in other areas of Czechoslovakia too (Moravia, s. Slovakia).

Aphidofauna of hops is rather poor. It includes a dioecious species namely — *Phorodon humuli* Schrk. It is attacked by three parasite species — *Ephedrus plagiator* (Nees), *Ephedrus persicae* Frog. and *Trioxys humuli* Mackauer. The first mentioned species was ascertained to infest the aphids on primary host plants *(Prunus)* only, *Trioxys humuli* Mackauer parasitized the aphid on hops only, while *Ephedrus persicae* Frog. was found to be a parasite of the aphid both on primary and secondary host plants. Nevertheless, due to the character of habitat it is believed the parasite complex will be identical both on primary and secondary host plants.

The collected material of parasites is not numerous in order to ascertain its foci. Because of the character of the community which is perennial, hop gardens represent probably temporary foci only.

It is difficult to evaluate the effectiveness of *Trioxys humuli* Mack. and other parasites of *Phorodon humuli* Schrk. as the material collected is rather poor.

13. Meadows (Plates XVII, XVIII, XIX, Figs. 33, 35, 36, 38, 39)

This is a type of widely distributed habitats in Czechoslovakia. In this type of habitat we include, because of the composition of parasite fauna, also the habitats of more or less natural "steppe" type, i.e. steppe preservations, meadows, etc. and their rests (pathways, balks, etc.) that occur in today's cultivated landscape.

The aphidofauna inhabiting these habitats is in a number of cases more differentiated than in cases of parasites that are, with some exceptions, more eurytopic than their hosts.

The typical parasite complexes may be recognized in the mentioned types of habitats:

The most typical and original steppe complex form parasites of the *Dactynotini*, being rather numerous in steppe preservations and steppe districts. They are: *Aphidius absinthii* Marsh., *Praon absinthii* (Bignell), both being parasites of *Macrosiphoniella*-species, *Aphidius funebris* Mack., *Praon dorsale* (Hal.), parasites of *Dactynotus*-species, *Aphidius phalangomyzi* Starý, a parasite of *Phalangomyzus*-species, *Trioxys centaureae* (Hal.) and *Ephedrus campestris* Starý, parasites of *Macrosiphoniella* and *Dactynotus*-species. The last member of this complex, *Trioxys pannonicus* Starý, which is a specialized parasite of *Titanosiphon artemisiae* (Koch), is more attached to the true steppe habitats occurring in southern Moravia, southern and eastern Slovakia.

Praon exoletum (Nees) is a typical specialized parasite of the *Therioaphis*-species that live on wild and cultivated leguminous plants.

Aphidius ervi Hal. is a parasite of *Acyrthosiphon*-species, which occur on wild and cultivated plants. It attacks also aphid species that live in more humid habitats like some kinds of waste places, where it attacks *Microlophium evansi* Theo. on *Urtica*.

A closely related species, *Aphidius mirotarsi* Starý, occurs in more or less typical steppe habitats, attacking *Mirotarsus cyparissae* (Koch) on *Euphorbia*-species.

Aphidius tanacetarius Mack. is a specialized parasite of *Metopeurum fuscoviride* Stroyan that occurs on *Tanacetum vulgare*, and *Microsiphum millefolii* Wahlg., which lives on *Achillea millefolium*.

Ephedrus nacheri Quilis M. P. attacks some gall-producing or leaf-curling aphid species *(Cryptomyzus, Hayhurstia)*.

Diaeretiella rapae (M'Int.) is a typical member that occurs in steppe habitats, a parasite of a number of Myzine aphids that live both on cultivated and wild plant species. *(Myzodes, Hayhurstia, Brevicoryne)*.

In steppe places, mostly in the true steppe, *Lysiphlebus arvicola* Starý may be found, a parasite of *Sipha*-species that occur on various grasses. This species is, like parasites of *Titanosiphon artemisiae* Bckt., a typical representative of the steppe fauna.

A number of root aphids is attacked by *Paralipsis enervis* (Nees). This parasite occurs also in steppe districts, especially on loessal soils, which are rather suitable for the occurrence of various root aphid species.

A number of aphids living on various grasses *(Sitobium, Metopolophium, Rhopalosiphum)* are attacked by *Aphidius avenae* Hal. and *Aphidius pascuorum* Marsh.

Coloradoa-species, occurring on *Achillea, Artemisia,* etc. are attacked by *Lysaphidus arvensis* Starý.

Lysaphidus erysimi Starý is a parasite of the leaf-curling aphid species *(Lipaphis, Pseudobrevicoryne)*.

Lysiphlebus fritzmuelleri Mack. is a specialized parasite of *Aphis craccae* L.

A number of Myzine aphids are attacked by *Aphidius matricariae* Hal. and *Aphidius picipes* (Nees).

Besides *Aphidius matricariae* Hal., which attacks Myzine aphids on *Galium* too, with these aphids *Trioxys parauctus* Starý is also associated. Other aphids occurring here *(Aphis galii-scabri* Schrk.) are attacked by specialized *Trioxys glaber* Starý.

The most common and typical representatives of the steppe fauna, which are distributed in all the habitats of the steppe type, are *Lysiphlebus fabarum* (Marsh.), *Lipolexis gracilis* Förster, *Trioxys acalephae* (Marsh.), being main parasites of *Aphis*-species and attacking also a number of other aphid groups.

The relation of the meadows s. lat. to the environmental habitats is rather important as they represent for the main part a source of parasite fauna which spreads from here to cultivated areas.

As the meadow type habitats represent at least several year communities, they include chronical foci of aphid parasites, both economically important and indifferent. From this viewpoint it is also necessary to evaluate the parasite fauna.

All the parasites of the *Dactynotini* are representatives of the indifferent parasites, although they are rather numerous. *Aphidius tanacetarius* Mack, *Ephedrus nacheri* Quilis, *Lysaphidus arvensis* Starý, *Lysaphidus erysimi* Starý, *Lysiphlebus fritzmuelleri* Mack., *Trioxys parauctus* Starý, *Trioxys glaber* Starý are similar cases. On the contrary, *Aphidius ervi* Hal., *Diaeretiella rapae* (M'Int.), *Lysiphlebus arvicola* Starý, *Paralipsis enervis* (Nees), *Aphidius avenae* Hal., *Aphidius pascuorum* (Marsh.), *Aphidius matricariae* Hal., *Aphidius picipes* (Nees), *Lysiphlebus fabarum* (Marsh.), *Lipolexis gracilis* Förster and *Trioxys acalephae* (Marsh.) are parasites of both economically indifferent and pest aphids so that their occurrence in habitats is important.

14. Ponds, Rushes

Due to the composition of parasite fauna this type of habitat may be regarded as an intermediate type. *Phragmites communis* represents a secondary host plant of *Hyalopterus pruni,* which is parasitized here by *Praon volucre* (Hal.), *Typha angustifolia* is a host plant of the monoecious aphid species *Schizaphis scirpi* (Kittel). This aphid is attacked by *Ephedrus plagiator* (Nees) and by *Diaeretiella rapae* (M'Int.).

172

The mentioned habitats have no specialized parasite fauna. On the contrary, because of the intermediate character of these habitats, there occur both representatives of the forest [*Ephedrus plagiator* (Nees), *Praon volucre* (Hal.)] or steppe type [*Diaeretiella rapae* (M'Int.)] habitats fauna.

15. Waste places (ruderal places)

In this type of habitat we include places which developed due to the activity of man, they are however not cared for by man but serve as places for storage of litter, wastes, etc. Old ruins of houses may be also included in this type of habitat.

Such habitats mentioned are inhabited by a number of typical weed plant species (*Chenopodium, Atriplex, Cirsium, Lappa, Carduus, Artemisia, Achillea*, etc.) with which corresponding aphid fauna is associated. For the main part, there occur the same aphid species and parasites as in the meadows. But, as there occur plants of a few species usually in a big quantity, the aphids and parasites are rather numerous in such places.

A part of parasite complexes inhabiting waste places is clearly indifferent. It includes parasites of the *Dactynotini* occurring on *Artemisia, Achillea, Carduus*, etc.

On the other hand, on a number of weed plants occur various aphid pests, which are parasitized by effective parasite species like *Lysiphlebus fabarum* (Marsh.) and *Lipolexis gracilis* Förster.

Because of the character of the plant community the waste places contain temporary foci of parasites.

As for the economic significance of foci of aphid parasites in waste places it is necessary to know under what conditions and when the pest aphids occur here. Usually, similar waste places represent sources of *Aphis fabae* Scop. and *Brachycaudus cardui* L. during summer. Both species occur here on secondary host plants and do not spread from here to cultivated crops (except possible secondary outbreaks of aphids on sugar beet, etc.). In the autumn only they spread to the primary host plants. As the aphids are usually heavily parasitized here by parasites, their existence in waste places may be kept as economically indifferent as only a small number migrate from here, while they augment the parasites in field type habitats. Nevertheless, also from this viewpoint it would be advantageous to prefer also in waste places the existence of plants, which are attacked by economically indifferent aphid species that include alternative hosts of economically valuable parasites of pest aphids.

16. Forage crop field (Plate XVIII, Fig. 37)

Forage crop fields are widely distributed both in the submountain districts and in the lowlands in Czechoslovakia. They represent mostly perennial communities (red-clover, alfalfa).

They are inhabited by pest aphid species, mainly *Acyrthosiphon pisum* (Harris), to a lesser degree by *Therioaphis* sp. and *Aphis craccivora* (Koch). *Acyrthosiphon pisum* (Harris) is attacked, similarly as in meadows, by *Aphidius ervi* Hal., *Therioaphis* sp. is parasitized by *Praon exoletum* (Nees), *Aphis craccivora* (Koch), for which alfalfa represents a secondary host plant, is attacked by *Lysiphlebus fabarum* (Marsh.) and *Lipolexis gracilis* Förster.

Because of the perennial character forage crop fields, the alfalfa fields namely, include chronical foci of parasites, which spread from there to other habitats.

Because of the life cycle of *Acyrthosiphon pisum* (Harris) and *Therioaphis* sp. and the character of the community, the parasites are able to reach high effectiveness, heavily attacking the mentioned aphid species.

17. Cereal crop field

Cereal crop fields are widely distributed in Czechoslovakia, in submountain districts *(Avena sativa)* and lowlands.

They represent annual crops. Their aphidofauna is composed mainly of dioecious aphid species (*Sitobium avenae* Kalt., *Rhopalosiphum*, *Metopolophium*, etc.) for which cereal crops represent secondary host plants, or monoecious aphid species (*Sitobium granarium* Kirby, *Sipha maydis* Pass, etc.). According to our observations the dioecious aphid occurrence on corn fields is the most common.

The parasites of the mentioned aphid species are represented mainly by the species *Aphidius avenae* Hal., to a lesser degree by *Aphidius pascuorum* Marsh., *Ephedrus plagiator* (Nees) and *Praon volucre* (Hal.). The last two mentioned species are typical for forest type habitats, from which they can spread and occur also in fields.

Because of the character of the community, cereal crop fields include temporary foci only. For this reason the occurrence of chronical parasite foci in the environment of these fields is important.

The economic significance of the above mentioned parasites is valuable as they are sometimes able to heavily infest the cereal crop aphid pests.

18. Leguminous crop field

Leguminous crops are grown in various districts of Czechoslovakia.

The aphidofauna is composed of the common aphid species, *Acyrthosiphon pisum* (Harris) and *Aphis craccivora* (Koch), which are parasitized by *Aphidius ervi* Hal. and *Lysiphlebus fabarum* (Marsh.).

Because of the annual character of leguminous crop fields they include foci of a temporary character only. For this reason the occurrence of chronic parasite foci in the neighbourhood (or perennial crops like alfalfa in case of *Aphidius ervi* Hal.) is important.

174

19. Oil crop field

Oil crops are grown in various places in Czechoslovakia, mostly in the lowlands.
The main part form the *Brassica*-species and *Helianthus annuus*, both being annual crops.

The *Brassica*-species are attacked by *Brevicoryne brassicae* (L.), sunflower by *Aphis fabae* Scop. and *Brachycaudus helichrysi* (Kalt.), *Brevicoryne brassicae* (L.) is parasitized by *Diaeretiella rapae* (M'Int.), *Aphis fabae* Scop. by *Lysiphlebus fabarum* (Marsh.).

As oil crops represent annual crops they may include temporary foci of parasites only. *Diaeretiella rapae* (M'Int.) spreads here from the neighbouring habitats, where it attacks *Brevicoryne*-aphids on various *Brassicaceae* weeds, and other economically indifferent aphids *(Hayhurstia)*. *Lysiphlebus fabarum* (Marsh.) is a similar case, attacking however other aphid species.

According to our observations, the infestation of *Brevicoryne brassicae* (L.) by *Diaeretiella rapae* (M'Int.) never broke down the aphid outbreak. As the *Brassicaceae* weeds represent also sources of *Brevicoryne* pest aphids, the occurrence in the neighbourhood of *Chenopodium* that is attacked by *Hayhurstia atriplicis* (Geoffr.) as the alternative host of *Diaeretiella rapae* (M'Int.) is suitable for the augmentation of parasites.

20. Vegetable crop field

Vegetable crops are grown in various lowland districts of Czechoslovakia.

The parasite complexes consist of members that occur in meadows. The aphidofauna is represented both by the monoecious and dioecious aphid species.

Diaretiella rapae (M'Int.) attacks *Brevicoryne brassicae* (L.) on *Brassica oleracea* var.

Aphidius salicis Hal. and *Trioxys brevicornis* (Hal.) parasitize *Semiaphis dauci* (F.) on carrots.

Because of the annual character of the vegetable crops their field may include temporary foci of parasites only. For the occurrence of parasites the existence of chronical foci in the environment is important.

The research of parasites of vegetable crop infesting aphids has been very poor in Czechoslovakia till now and for this reason no data on the economic importance of parasites can be given.

21. Potato field (Plate XIX, Fig. 40)

Potato is grown mostly in submountain districts in Czechoslovakia.

Potato plants are attacked by a number of aphids, from which *Myzodes persicae* (Sulz.) was found to be parasitized by *Diaeretiella rapae* (M'Int.).

175

Potato field represents an annual crop, including temporary foci of parasites only. For the occurrence of *Diaretiella rapae* (M'Int.) in the field the presence of chronic foci is necessary, such as waste places covered by *Chenopodium* spp. as the host plant of *Hayhurstia atriplicis* (F.), an alternative host of *Diaeretiella rapae* (M'Int.). Similarly, the occurrence of *Chenopodium* sp. in potato field, although it is a weed plant, is suitable for the augmentation of *Diaeretiella rapae* (M'Int.) effectiveness, which attacks also *Myzodes persicae* (Sulz.) on potatoes.

According to our observations *Myzodes persicae* (Sulz.) was heavily parasitized on potato by *Diaeretiella rapae* (M'Int.), although this parasite is not so effective in the case of *Brevicoryne brassicae* (L.) control. This is probably caused by too numerous and dense colonies of *Brevicoryne* in comparison with *Myzodes*.

22. Sugar beet field (Plate XVII, Fig. 32)

Sugar beet is commonly grown in lowland districts of Czechoslovakia, in central and eastern Bohemia, central and southern Moravia and in southern Slovakia.

Sugar beet is attacked mostly by *Aphis fabae* Scop., to a lesser degree by *Myzodes persicae* (Sulz.). In both cases sugar beet represents a secondary host plant of the mentioned aphid species. *Aphis fabae* Scop. is parasitized by *Lysiphlebus fabarum* (Marsh.), *Lipolexis gracilis* Förster; *Myzodes persicae* (Sulz.) is attacked by *Diaeretiella rapae* (M'Int.).

Sugar beet fields represent an annual community (rarely biennal), including temporary foci of parasites only. The parasites must spread there from the neighbouring habitats, where chronic foci exist. For this reason the presence of various balks, pathways, ditches, etc. in the environment of sugar beet fields is valuable for parasite occurrence.

Because of their spreading possibilities the parasites are hardly able to prevent an aphid outbreak. They are sometimes effective after the injury to plant is made by aphids. For this reason it seems they may play only a secondary role in the integrated aphid control. The occurrence of *Chenopodium*-plants as weeds (see *Solanum*) in sugar beet fields is not recommended in the case of sugar beet as *Chenopodium* represents also the host plant of *Aphis fabae* Scop. and a source of a possible second aphid outbreak.

23. Tobacco field

Tobacco is grown only in southern districts in Czechoslovakia.

It is attacked by *Myzodes persicae* Sulz., as the secondary host plant of this aphid.

As tobacco is an annual crop, it is believed there will be a similar situation as to the parasites as in a potato field.

176

Myzodes persicae Sulz., because of the lack of reared material, it is expected to be parasitized by *Diaeretiella rapae* (M'Int.).

Because of the lack of material of parasites no data on the economic significance of parasite effectiveness can be given.

A scheme of parasite focus research

The research of foci of parasites is one of the basic facilities for the research of natural limitation of aphids by parasites and it is necessary for the future of integrated control. There are in general two directions in the research of foci of aphid parasites:

1. Theoretical research of foci, dealing with classification of foci, relations between aphids and parasites in field conditions, spread of aphids and parasites from foci to cultural areas, factors influencing the effectiveness of parasites, etc. This research needs a relatively high level of specialised work and a good rate of aphidological and entomophagological knowledge. It has approximately the following points:

a) The basic types of landscapes of the studied area are studied and the main types of habitats selected.

The character of habitat, plant community, qualitative composition of the aphido-fauna and its parasites are studied first, on this basis the classification and research of various intermediary habitats is made.

b) After the preliminary research of the aphidofauna and its parasites the relations of aphids and parasites in different types of habitats in relations to the peculiarities of their life cycle are studied. In this period of work all the obtainable colonies of aphids are collected to ascertain the percentage of occurrence of various parasite species in different types of habitats and their food specificity.

c) The analysis of habitats with regard to the classification of foci is made.

d) Research of parasite species in different types of foci (chronic, temporary) with regard to their seasonal occurrence, peculiarities of the life cycle, etc.

e) Selection of the main aphid complexes with regard to the pest species and their parasite complexes and elaboration of a further scheme of work concerning the given problematics.

f) Elaboration of arrangements enabling the protection of the existing foci and efforts for creation of new foci.

2. Applied research of foci with regard to a certain aphid species and all the food chain of its parasites. This kind of research presupposes at least a basic knowledge of theoretical research (point 1) and such elaboration of aphid fauna and fauna of parasites especially, when their main ecological characteristics (hosts, food specificity, occurrence, habitats, etc.) are known. This direction of research has the following points:

a) Life cycle of the given aphid species and its occurrence in various types of habitats.

b) Sampling of all obtainable colonies of the given aphid species in different kinds of habitats in which it occurs.

c) Analysis of parasites.

d) Selection of the most common and effective species and their classification according to the data obtained in the theoretical research.

e) Research of natural limitation of aphids by parasites (effectiveness, spread, occurrence, bionomics and ecology).

f) Further efforts on the augmentation of natural enemies for integrated control purpose (see chapter XIV).

The significance and protection of foci of aphid parasites

Research on the occurrence of aphid parasites in nature has shown that the foci of parasites in nature are of different types. According to the character of habitats and agricultural activities a special situation has developed due to the crop rotation, another annual crop growing every year in the same field. This feature has, in connection with the life-cycle of aphids, too, the result that in annual crops the parasites may hardly influence the occurrence of aphids in larger areas at the initial stage, especially if there does not exist a sufficient number of suitable habitats in the neighbourhood, where a number of chronic foci of parasites may be found. Even in the latter case the parasites spread only gradually all over the cultural area from the chronic foci.

The parasites may be theoretically, and often practically, too, effective economically particularly on perennials, mostly on forage crops, on which monoecious aphid species occur during the whole season, and where chronic foci of parasites may gradually originate.

A similar situation is in forest type habitats, which is economically important namely in orchards.

A generally known case may be mentioned in this connection, that of *Aphis fabae* occurring on various weeds in field boundaries, pathways, etc. The aphid is usually heavily destroyed in summer by parasites so that a lower number of aphids may migrate to primary host plants and serve as a source of a possible outbreak in the following year.

Our studies have shown that foci of parasites are of different types. The protection of a focus must be based on a detailed analysis of the host plants, host aphids and complexes of parasites of the respective foci, by which the structure and type of a given focus becomes established. Indifferent foci of parasites may not be excluded from this viewpoint. Similarly, we shall not protect various weeds if other suitable plants do the same service for our purposes: E.g. we shall not protect *Cirsium* plants in fields, as it is a weed plant and a secondary host plant of *Aphis fabae*, although the aphid here is usually heavily infested by *Lysiphlebus fabarum* (Marsh.), as this parasite may attack other economically indifferent aphids in the same type of habitat, too.

The further direction of work seems to include the influencing of covercrop in suitable habitats and augment the aphid parasites through the corresponding aphidofauna in this way.

XII.

A review
of pest aphids
in agriculture and forestry
and their
parasites

The purpose of this review is to present the general features of the life cycle of pest aphids and a list of their parasites on various crops. Further records on parasites may be found in chapter IV. Only such aphid species from which parasites were bred in Czechoslovakia are listed. In some cases of aphid pests from which parasites have not been bred in Czechoslovakia till now but which are known to the author from the neighbouring countries, such a parasitic occurrence is mentioned as probable.

Aphids-virus vectors are marked by an asterisk(*).

A. Agriculture

Cereal grains

Triticum vulgare, Hordeum distichon, Avena sativa, Secale cereale.

Anoecia spp. (*corni*, etc.). Dioecious. Primary host plant — *Cornus*, in forest type and intermediary habitats, attacked by *Lipolexis gracilis* Förster. Secondary host plant — on roots of grasses; attacked by *Paralipsis enervis* (Nees). Economic importance small.

Metopolophium dirhodum (Walk.). Dioecious. Primary host plant — *Rosa, Fragaria*, in forest and intermediary habitats, attacked probably by *Ephedrus plagiator* (Nees) and similar parasite species as in the case of *Sitobium*-species. Secondary host plants — *Gramineae* (*Avena sativa* and wild grasses), attacked probably by the same parasite species as *Sitobium* spp. Economic importance small.

Rhopalosiphum padi (L.). Dioecious. Primary host plant — *Padus racemosa*, in forest type habitats, attacked by *Ephedrus persicae* Frog., *Ephedrus plagiator* (Nees), *Praon abjectum* (Hal.), *Trioxys angelicae* (Hal.). Secondary host plants — on corn and wild grasses, attacked by *Ephedrus plagiator* (Nees) and probably by *Aphidius avenae* Hal. Occurrence common.

180

Sipha maydis (Pass.). Monoecious, occurring on various grasses, in steppe habitats, attacked by *Lysiphlebus arvicola* Starý. Occurrence common namely on wild grasses.

Sitobium avenae (Fabr.). Dioecious. Primary host plants — *Rosa. Rubus, Fragaria,* in forest and intermediary habitats, attacked by *Ephedrus plagiator* (Nees). Secondary host plant — *Gramineae,* attacked by *Aphidius avenae* Hal., *Aphidius pascuorum* Marsh., *Ephedrus plagiator* (Nees), *Lysiphlebus fabarum* (Marsh.).

Sitobium granarium (Kirby). Monoecious. In steppe habitats on *Gramineae,* attacked probably by *Ephedrus plagiator* (Nees), and other species as in the preceding species.

Both *Sitobium*-species occur often as pest aphids on corn.

Forage crops

Trifolium spp.

Acyrthosiphon pisum (Harris). Monoecious, in steppe type habitats, attacked by *Aphidius ervi* Hal., *Praon dorsale* (Hal.). Rather important pest aphid.

Medicago sativa

Acyrthosiphon pisum (Harris). Monoecious, in steppe habitats, attacked by *Aphidius ervi* Hal. and *Praon dorsale* (Hal.). A very important pest aphid, namely in southern districts.

Therioaphis sp. Monoecious, in steppe habitats, attacked by *Praon exoletum* (Nees). Common in southern districts namely, occurring sometimes as pest aphid.

Aphis craccivora (Koch). Monoecious, in steppe habitats, attacked by *Lipolexis gracilis* Förster, *Lysiphlebus fabarum* (Marsh.). Occurrence occasional.

Leguminous crops

Vicia faba et spp.

Acyrthosiphon pisum (Harris). Monoecious, in steppe habitats, attacked by *Aphidius ervi* Hal. Occurrence occasional.

Aphis craccae (Koch). Monoecious, in steppe habitats, attacked by *Ephedrus plagiator* (Nees), *Lysiphlebus fritzmuelleri* Mack., *Trioxys acalephae* (Marsh.). Occurrence on economically important *Vicia*-species occasional.

Aphis fabae Scop. Dioecious (see *Beta vulgaris*). Attacked by *Lysiphlebus fabarum* (Marsh.). Occurrence — occasionally as serious pest aphid.

Megoura viciae Bckt. Monoecious, in steppe and intermediate habitats, attacked by *Aphidius megourae* Starý and *Praon dorsale* (Hal.). Occurrence occasional.

Oil crops

Brassica napus var. *oleifera*

Brevicoryne brassicae (L.) Monoecious, on *Cruciferae* (wild and cultured) in steppe habitats, attacked by *Diaretiella rapae* (M'Int.). Important pest aphid.

Brassica rapa var. *oleifera*

Brevicoryne brassicae (L.). see above, attacked by *Diaeretiella rapae* (M'Int.). Important pest aphid.

Papaver somniferum

Aphis fabae Scop. see *Beta vulgaris;* attacked probably by *Lysiphlebus fabarum* (Marsh.). Important pest aphid.

Myzodes persicae (Sulz.). Dioecious (see *Prunus persica*). Attacked by *Aphidius picipes* (Nees). *Trioxys angelicae* (Hal.). Occasionally pest aphid.

Helianthus annuus

Aphis fabae Scop. (see *Beta vulgaris*). Occurrence occasional.

Vegetable crops

Lactuca sativa

Macrosiphum euphorbiae Theo. Widely polyphagous, in steppe habitats occurring species, attacked by *Praon volucre* (Hal.).

Nasonovia ribisnigri (Mosl.). Dioecious. Primary host plant — *Ribes,* in forest and intermediate habitats. Secondary host plants — *Hieracium, Crepis, Cichorium, Lactuca,* etc. in forest and intermediate habitats, attacked by *Aphidius hieraciorum* Starý and *Monoctonus crepidis* (Haliday). Occasional occurrence.

Brassica oleracea et var.

Brevicoryne brassicae (L.). Monoecious (see *Brassica* sp.). Attacked by *Diaeretiella rapae* (M'Int.) and *Praon volucre* (Hal.). Important pest aphid.

Daucus carota

Dysaphis crataegi (Kalt.). Dioecious. Primary host plants — *Crataegus,* or *Malus,* in forest type habitats, attacked probably by *Ephedrus persicae* Frog. and *Ephedrus plagiator* (Nees). Secondary host plants — on roots of *Daucus carota,* attacked by *Paralipsis enervis* (Nees). Occurrence occasional.

Semiaphis dauci (F.). Monoecious. In steppe habitats, attacked by *Aphidius salicis* Hal. Occasional occurrence as pest aphid.

Potato

Aphis nasturtii (Kalt.). Dioecious. Primary host plants — *Rhamnus cathartica,* in forest and intermediate habitats, attacked by *Ephedrus plagiator* (Nees). Secondary host plants — cruciferous plants, in steppe habitats. Occurrence common.

182

Myzodes persicae (Sulz.). Dioecious (see *Prunus persica*). Attacked by *Aphidius picipes* (Nees), *Diaeretiella rapae* (M'Int.). Common pest aphid.

Macrosiphum euphorbiae Theo. (see *Lactuca sativa*).

Sugar beet

Aphis fabae Scop. Dioecious. Primary host plant — *Euonymus,* in forest and intermediate habitats, attacked by *Ephedrus persicae* Frog., *Ephedrus plagiator* (Nees), *Praon abjectum* (Hal.), *Trioxys angelicae* (Hal.). Secondary host plants — *Beta vulgaris, Cirsium* spp., *Arctium* spp., *Carduus* spp., etc. in steppe habitats, attacked by *Lysiphlebus fabarum* (Marsh.), *Lipolexis gracilis* Förster, *Trioxys acalephae* (Marsh.). Rather important pest aphid.

Myzodes persicae (Sulz.) (see *Prunus persica*). Attacked by *Diaeretiella rapae* (M'Int.). Important pest aphid-virus vector.

Hop

Phorodon humuli (Schrk.). Dioecious. Primary host plant — *Prunus domestica, Prunus spinosa,* in forest and intermediate habitats, attacked by *Ephedrus persicae* Frog., *Ephedrus plagiator* (Nees). Secondary host plants — *Humulus lupulus,* in intermediate habitats, attacked by *Trioxys humuli* Mackauer. Important pest aphid.

Tobacco

Myzodes persicae (Sulz.) (see *Prunus persica*). Attacked probably by *Diaeretiella rapae* (M'Int.) Important pest aphid namely as virus vector.

Fruit trees and shrubs

Ribes spp.

Cryptomyzus ribis (L.). Dioecious. Primary host plant — *Ribes rubrum,* in forest and intermediate habitats, attacked by *Aphidius ribis* Hal. Secondary host plant — *Stachys, Lamium,* in steppe type habitats. Important pest aphid.

Hyperomyzus lactucae (L.). Dioecious. Primary host plant — *Ribes nigrum,* in forest and intermediate habitats, attacked by *Ephedrus plagiator* (Nees). Secondary host plant — *Sonchus,* etc. in steppe habitats, attacked by *Aphidius sonchi* Marsh., *Lysiphlebus fabarum* (Marsh.), *Praon volucre* (Hal.). Occasional pest aphid.

Nasonovia ribisnigri (Mosl.). (see *Lactuca*).

Schizoneura ulmi (L.). Dioecious. Primary host plant — *Ulmus*, in forest type habitats, attacked by *Areopraon lepelleyi* (Wat.). Secondary host plant — on roots of *Ribes*, parasites unknown.

Pirus communis

Dysaphis spp. Dioecious. Primary host plants — *Pirus*, in forest and intermediate habitats, attacked by *Ephedrus persicae* Frog. and *Ephedrus plagiator* (Nees). Secondary host plants — various, depending on the species, attacked by *Lysiphlebus fabarum* (Marsh.) and (probably) by *Paralipsis enervis* (Nees). Occurrence occasional as pest aphid.

Malus silvestris

Aphis pomi Deg. Monoecious. In forest and intermediate habitats on *Crataegus*, *Sorbus*, *Malus*, *Cotoneaster*, attacked by *Ephedrus plagiator* (Nees), *Lysiphlebus fabarum* (Marsh.) occasionally, *Trioxys angelicae* (Hal.). Rather important pest aphid namely on young apple trees.

Dysaphis spp. (*crataegi*, *devecta*, etc.). Dioecious. Primary hosts plants — *Malus*, *Crataegus*, in forest and intermediate habitats, attacked by *Ephedrus persicae* Frog., *Ephedrus plagiator* (Nees), *Trioxys angelicae* (Hal.), *Praon volucre* (Hal.). Secondary host plants — on roots of various plants in steppe habitats, attacked by *Lysiphlebus fabarum* (Marsh.) and *Paralipsis enervis* (Nees). Occurrence often as serious pest aphid.

Allocotaphis quaestionis Börn. Dioecious. Primary host plant — *Malus*, in forest and intermediate habitats, attacked by *Ephedrus persicae* Frog. Secondary host plant — unknown. Occurrence occasional as pest aphid in a similar way as *Dysaphis*-species.

Rubus idaeus

Aphis idaei v. d. G. Monoecious, in forest and intermediate habitats, attacked by *Ephedrus persicae* Frog. and *Ephedrus plagiator* (Nees). Occurrence as pest aphid on cultivated *Rubus-species*.

Nectarosiphon rubi (Theo.). Monoecious, in forest and intermediate habitats, attacked (probably) by *Aphidius rubi* Starý.

Fragaria spp.

Nectarosiphon rubi (Theo.) (see *Rubus*).

Prunus cerasus

Myzus cerasi (F.). Dioecious. Primary host plant — *Prunus avium*, *Prunus cerasus*, in forest and intermediate habitats, attacked by *Aphidius matricariae* Hal., *Ephedrus cerasicola* Starý, *Ephedrus persicae* Frog., *Ephedrus plagiator* (Nees), *Lipolexis gracilis* Först. Secondary host plant — *Galium* spp., in steppe habitats, parasites unknown. Rather serious and important pest aphid.

Prunus domestica

Brachycaudus cardui (L.) Dioecious. Primary host plants — *Prunus domestica*, in forest and intermediate habitats, attacked by *Ephedrus plagiator* (Nees). Secondary

184

host plants — *Carduus, Arctium,* etc. in steppe habitats, attacked by *Lipolexis gracilis* Först., *Lysiphlebus fabarum* (Marsh.). Occasional occurrence as pest aphid.

*Brachycaudus helichrysi (Kalt.). Dioecious. Primary host plant — *Prunus domestica, Prunus spinosa,* etc. in forest and intermediate habitats, attacked by the same parasite species as in the preceding species. Secondary host plants — *Arctium, Artemisia, Achillea, Centaurea, Carduus,* etc. in steppe habitats, attacked by the same parasites as in the preceding species. Rather serious pest aphid on young *Prunus armeniaca* and *Prunus persica* trees.

*Hyalopterus pruni (Geoffr.). Dioecious. Primary host plant — *Prunus domestica, Prunus spinosa, Prunus persica, Prunus armeniaca,* in forest and intermediate habitats, attacked by *Ephedrus plagiator* (Nees) and *Praon volucre* (Hal.). Secondary host plant — *Phragmites communis,* in intermediate habitats, attacked by the same parasite species as on primary host plants. Serious pest aphid.

*Phorodon humuli Schrk. (see *Humulus lupulus*).

*Rhopalosiphum nymphaeae (L.). Dioecious. Primary host plant — *Prunus armeniaca, Prunus persica, Prunus domestica, Prunus spinosa,* in forest and intermediate habitats, parasites unknown. Secondary host plants — *Hydrochaeris morsus ranae, Meniathes trifoliata, Nymphaea* sp., *Nuphar* sp., etc. in intermediate habitats, attacked by *Praon necans* Mack. Occurrence occasional.

Prunus armeniaca

*Hyalopterus pruni (Geoffr.) (see *Prunus domestica*).

*Rhopalosiphum nymphaeae (L.) (see *Prunus domestica*).

Prunus persica

*Brachycaudus helichrysi (Kalt.) (see *Prunus domestica*).

*Hyalopterus pruni (Geoffr.) (see *Prunus domestica*).

*Myzodes persicae (Sulz.). Monoecious or facultatively dioecious, in intermediate and steppe habitats (see *Beta vulgaris, Solanum tuberosum*).

*Rhopalosiphum nymphaeae (L.) (see *Prunus domestica*).

Juglans regia

Callaphis juglandis (Goetze). Monoecious, in forest and intermediate habitats — parasites unknown although numerous samples were reared.

Chromaphis juglandicola (Kalt.). Monoecious, in forest and intermediate habitats, attacked by *Trioxys pallidus* (Hal.). Occurrence common, economic importance small.

Corylus avellana

Myzocallis coryli (Goetze). Monoecious, in forest and intermediate habitats, attacked by *Ephedrus plagiator* (Nees). Occurrence occasional, economic importance small.

Vitis vinifera

Viteus vitifolii (Fitch). Parasites unknown.

B. Forestry

Forest trees, shady trees, ornamental trees

Juniperus communis

*Cupressobium·juniperi (Deg.). In forest and intermediate habitats, attacked by *Pauesia juniperorum* (Starý) and *Pauesia cupressobii* (Starý). Occurrence common, economic importance small.

Picea excelsa

Cinara spp. In forest habiṭats, attacked by various *Pauesia* species. Occurrence occasional.

Lachniella costata Zett. In forest habitats, attacked by (probably) various *Pauesia*-species.

Abies alba

Buchneria pectinatae (Nördl.). In forest habitats, attacked by *Pauesia infulata* (Hal.). Occurrence occasional, sometimes as pest aphid.

Todolachnus abieticola (Chol.). In forest habitats, attacked by *Pauesia grossa* (Fahr.). Occurrence-occasionally as pest aphid.

Elatobium abietinum (Walk.). In forest habitats, attacked probably by *Lysaphidus schimitscheki* Starý. Serious pest aphid.

Larix decidua

Cinara spp. ln forest habitats, attacked by various *Pauesia* species. Occurrence common.

Pinus silvestris, nigra, mugo, etc.

Protolachnus agilis (Kalt.). In forest and intermediate habitats, attacked by *Praon bicolor* Mack. and *Diaeretus leucopterus* (Hal.). Occurrence common.

Schizolachnus pineti (F.). In forest and intermediate habitats, attacked by *Pauesia unilachni* (Gah.). Occurrence common.

Cinara spp. In forest and intermediate habitats, attacked by various *Pauesia*-species, *Metaphidius aterrimus* (Fahr.). Occurrence common.

Salix ssp. (*alba, viminalis, caprea,* etc.)

*Aphis farinosa Gmel. Monoecious, in forest and intermediate habitats, attacked by *Ephedrus plagiator* (Nees), *Lysiphlebus ambiguus* (Hal.), *Praon abjectum* (Hal.), *Trioxys acalephae* (Marsh.), *Trioxys angelicae* (Hal.). Occurrence common.

*Pterocomma spp. Monoecious, in forest and intermediate habitats, attacked by *Aphidius pterocommae* Ashm. Occurrence occasional.

*Cavariella spp. Dioecious. Primary host plants — *Salix* spp., in forest and intermediate habitats, attacked by *Aphidius salicis* Hal. Secondary host plants — *Angelica, Daucus,* etc. in intermediate and steppe habitats, attacked by the same parasite species as on primary host plants. Occurrence occasional.

Populus spp. (*alba, tremula, nigra,* etc.).

186

Chaitophorus spp. Monoecious, in forest and intermediary habitats, attacked by *Lysiphlebus salicaphis* (Fitch). Occurrence common.

*Pterocomma spp. Monoecious, in forest and intermediate habitats, attacked by *Aphidius pterocommae* Ashm. Occurrence occasional.

Betula spp. *(verrucosa, pubescens)*

Calaphis spp. Monoecious, in forest and intermediate habitats, attacked by *Aphidius sicarius* Mack. Occurrence common.

Glyphina betulae (Kalt.). Monoecious, in forest and intermediate habitats, attacked probably by *Aphidius sicarius* Mack. Occurrence common.

Symydobius oblongus (v. Heyd.). Monoecious, in forest and intermediate habitats, attacked by *Aphidius sicarius* Mack.

Carpinus betulus

Myzocallis carpini (Koch). Monoecious, in forest habitats, attacked by *Praon flavinode* (Hal.) and *Trioxys pallidus* (Hal.). Occurrence common.

Fagus silvatica

Phyllaphis fagi (L.) Monoecious, in forest habitats, attacked by *Trioxys phyllaphidis* Mack. Occurrence common, often as serious pest aphid.

Quercus spp. *(pubescens, sessilis, robur,* etc.)

Stomaphis sp. Monoecious, in forest habitats, attacked by *Protaphidius wissmannii* (Ratz.). Occurrence rare.

Thelaxes dryophila (Schrk.). Monoecious, in forest habitats, attacked by *Lysiphlebus thelaxis* Starý. Occurrence common.

Tuberculoides annulatus (Htg.) Monoecious, in forest habitats, attacked by *Praon flavinode* (Hal.) and *Trioxys pallidus* (Hal.). Occurrence common.

Ulmus campestris et spp.

Tinocallis platani (Kalt.). Monoecious, in forest and intermediate habitats, attacked by *Praon flavinode* (Hal.) and *Trioxys hortorum* Starý. Occurrence common.

Tilia spp.

Eucallipterus tiliae (L.). Monoecious, in forest and intermediate habitats, attacked by *Praon flavinode* (Hal.), *Trioxys pallidus* (Hal.). Occurrence common, sometimes as pest aphid.

Sorbus aucuparia (et spp.)

Dysaphis sorbi (Kalt.) (and spp.). Dioecious. Primary host plant — *Sorbus,* in forest and intermediate habitats, attacked by *Ephedrus persicae* Frog., *Ephedrus plagiator* (Nees). Secondary host plant — *Campanula* spp., in steppe habitats, parasites unknown, probably *Paralipsis enervis* (Nees).

Acer spp. *(pseudoplatanus, platanoides, campestre,* etc.)

*Drepanosiphon platanoidis (Schrk.). Monoecious, in forest habitats, attacked by *Dyscritulus planiceps* (Marsh.), *Trioxys cirsii* (Curt.) and *Monoctonus pseudoplatani* (Marsh.). Occurrence common.

*Periphyllus villosus (Htg.). Monoecious, in forest and intermediate habitats, attacked by *Trioxys falcatus* Mack. Occurrence common.

Cornus sanguinea

Anoecia spp. Dioecious. Primary host plant — *Cornus* spp., in forest and intermediate habitats, attacked by *Lipolexis gracilis* Förster. Secondary host plants — on roots of various plants, attacked by *Paralipsis enervis* (Nees). Occurrence common.

Berberis vulgaris

Liosomaphis berberidis (Kalt.). Monoecious, in forest and intermediate habitats, attacked by *Aphidius hortensis* Marsh. and *Ephedrus plagiator* (Nees). Occurrence common.

Philadelphus coronarius

Aphis fabae Scop. (see *Euonymus europaea*).

Spiraea spp.

Aphis spiraephaga Müller. Monoecious, in forest and intermediate habitats, attacked by *Ephedrus plagiator* (Nees), *Trioxys acalephae* Marsh., *Trioxys angelicae* (Hal.). Occurrence common.

Rosa spp.

Sitobium avenae (Fabr.). Dioecious. Primary host plant — *Rosa* spp., in forest and intermediate habitats, attacked by *Ephedrus plagiator* (Nees). Secondary host plants — *Gramineae* (see grains). Occurrence common.

Macrosiphum rosae (L.). Dioecious, occasionally monoecious. Primary host plant — *Rosa* spp., in forest and intermediary habitats, attacked by *Aphidius rosae* Hal., *Ephedrus plagiator* (Nees), *Praon rosaecola* Starý, *Praon volucre* (Hal.). Secondary host plants — *Dipsacus, Scabiosa, Knautia*, in steppe and intermediate habitats, attacked probably by the same or similar species as on primary host plants. Occurrence common, often as serious pest aphid.

Viburnum spp.

Aphis viburni Scop.? Dioecious. In forest and intermediate habitats, attacked by *Praon abjectum* (Hal.) and *Trioxys angelicae* (Hal.).

Ceruraphis eriophori (Walk.). Dioecious. Primary host plant — *Viburnum* spp., in forest and intermediary habitats, attacked by *Ephedrus plagiator* (Nees), *Trioxys angelicae* (Hal.). Secondary host plant — *Carex*. Occurrence common.

Lonicera spp.

Hyadaphis mellifera (Hottes). Dioecious. Primary host plant — *Lonicera*, in forest and intermediary habitats, attacked by *Ephedrus persicae* Frog. Occurrence common.

Sambucus nigra

Aphis sambuci L. Dioecious. Primary host plant — *Sambucus*, in forest and intermediary habitats, attacked by *Praon abjectum* (Hal.) and *Trioxys angelicae* (Hal.). Secondary host plants — *Rumex*, etc. Occurrence common.

Ligustrum vulgare

Myzodes ligustri (Mosl.) Monoecious, in forest and intermediate habitats, attacked by *Ephedrus persicae* Frog. and *Monoctonus cerasi* (Marsh.). Occurence occasional.

Robinia pseudoacacia

188

Aphis craccivora (Koch). Monoecious, in various kinds of habitats, in forest and intermediate habitats attacked by *Praon abjectum* (Haliday). Occurrence common.

Euonymus europaea

Aphis evonymi F. Dioecious. Primary host plant — *Euonymus,* in forest and mainly in intermediate habitats, attacked probably by the same species as *Aphis fabae* Scop. Secondary host plants — *Solanum, Rumex,* etc., in steppe habitats, attacked by *Lysiphlebus fabarum* (Marsh.). Occurrence common.

Aphis fabae Scop. (see *Beta vulgaris*).

Aphis cognatella Jones. Monoecious (validity uncertain). In forest and intermediate habitats, attacked by the same species as *Aphis fabae* Scop.

Padus racemosa

Rhopalosiphum padi (L.). Dioecious. Primary host plant — *Padus,* in forest and intermediary habitats, attacked by *Ephedrus persicae* Frog., *Ephedrus plagiator* (Nees), *Praon abjectum* (Hal.), *Trioxys angelicae* (Hal.). Secondary host plants — grains (see *Triticum,* etc.). Occurrence common, sometimes as a serious pest.

Prunus mahaleb

Roepkea marchali (Börner).? Monoecious. In forest and intermediate habitats, attacked by *Ephedrus persicae* Frog. Occurrence common, often as serious pest aphid.

Crataegus spp.

Aphis pomi Deg. (see *Malus silvestris*).

Dysaphis spp. (see *Malus silvestris*).

Rhamnus cathartica

Aphis nasturtii (Kalt.). Dioecious. Primary host plant — *Rhamnus,* in forest and intermediate habitats, attacked by *Ephedrus plagiator* (Nees). Occurrence common.

Caragana arborescens

Acyrthosiphon caraganae Chol. Monoecious, in forest and intermediate habitats, attacked by *Aphidius caraganae* Starý, *Ephedrus plagiator* (Nees), *Praon volucre* (Hal.), *Toxares deltiger* (Hal.), *Trioxys angelicae* (Hal.). Occurrence common, often as a serious pest aphid.

XIII.

Host
and
parasite
catalogue

In this chapter only original records are included. Similarly, as in all the paper, only exactly identified material of parasites is listed, while taxonomically unclear species were omitted.

Because of the purpose of this paper it was not possible to include all the localities so that the preference of different parasite species as to the habitats and aphid host plants is not clear. This matter is dealt with in ecological characteristics of parasites, where the preference is generalized.

As to the aphid nomenclature, the classification of Börner (1952), with some exceptions, has been followed.

Acyrthosiphon caraganae (Chol.)
Aphidius caraganae Starý: On *Caragana arborescens*
Aphidius ervi Haliday: On *Colutea arborescens*
Ephedrus plagiator (Nees): On *Caragana arborescens*
Praon volucre (Haliday): On *Caragana arborescens*
Toxares deltiger (Haliday): On *Caragana arborescens*
Trioxys angelicae (Haliday): On *Caragana arborescens*

Acyrthosiphon pisum (Harris)
Aphidius ervi Haliday: On *Dorycnium herbaceum, Lathyrus* sp., *Medicago sativa, Medicago varia, Melilotus albus, Melilotus officinalis, Trifolium pratense, Trifolium montanum, Trifolium recuspinatum, Vicia cracca, Vicia faba, Vicia* sp.
Praon dorsale (Haliday): On *Medicago sativa, Trifolium pratense, Vicia cracca*

Acyrthosiphon spartii (Koch)
Aphidius ervi Haliday: On *Sarothamnus scoparius*
Ephedrus plagiator (Nees): On *Sarothamnus scoparius*

Acyrthosiphon superbum Börner
Aphidius ervi Haliday: On *Sesseli hypomarathum, Sesseli osseum*
Acyrthosiphon spp.

190

Aphidius ervi Haliday: On *Astragalus vesicarius, Oxytropis pilosa*
  Allocotaphis quaestionis (Börner)
Ephedrus persicae Froggatt: On *Malus silvestris*
  Amphorophora ampullata Bckt.
Aphidius lonicerae (Marshall): On *Dryopteris austriaca*
  Anoecia spp.
Lipolexis gracilis Förster: On *Cornus sanguinea*
Paralipsis enervis (Nees): On *Agropyrum repens*
  Aphidoidea spp.
Aphidius salicis Haliday: On *Conium maculatum*

  Aphis bupleuri (Börner)
Lipolexis gracilis Förster: On *Bupleurum falcatum*
Praon abjectum (Haliday): On *Bupleurum falcatum*
  Aphis cognatella Jones
Trioxys angelicae (Haliday): On *Euonymus europaea*

  Aphis craccae (L.)
Ephedrus plagiator (Nees): On *Vicia cracca*
Lysiphlebus fritzmuelleri Mackauer: On *Vicia sepium, Vicia cracca*
Trioxys acalephae (Marshall): On *Vicia* sp.

  Aphis craccivora (Koch)
Lipolexis gracilis Förster: On *Medicago sativa, Onobrychis sativa*
Lysiphlebus fabarum (Marshall): On *Medicago sativa, Onobrychis sativa, Trifolium*
  sp., *Vicia* sp.
Praon abjectum (Haliday): On *Robinia pseudoacacia*
Praon volucre (Haliday): On *Caragana arborescens*

  Aphis cystisorum (Htg.)
Lysiphlebus fabarum (Marshall): On *Laburnum anagyroides*
Trioxys acalephae (Marshall): On *Trifolium* sp., *Onobrychis sativa*
  Caragana arborescens, Robinia pseudoacacia
Trioxys angelicae (Haliday): On *Caragana arborescens*

  Aphis euphorbiae (Kalt.)
Lipolexis gracilis Förster: On *Euphorbia cyparissias*
Lysiphlebus fabarum (Marshall): On *Euphorbia cyparissias*
Trioxys acalephae (Marshall): On *Euphorbia cyparissias*

  Aphis evonymi F.
Lysiphlebus fabarum (Marshall): On *Solanum nigrum, Fagopyrum* sp.

  Aphis fabae Scop.
Ephedrus persicae Froggatt: On *Euonymus europaea, Matricaria inodora*
Ephedrus plagiator (Nees): On *Beta vulgaris, Borago officinalis, Cirsium* sp., *Cheno-*
  podium sp., *Euonymus europaea, Epipactis latifolia, Impatiens nolli-tangere,*
  Impatiens roepkei, Philadelphus coronarius

Lipolexis gracilis Förster: On *Beta vulgaris, Cirsium* sp., *Cirsium arvense, Centaurea cyanus*

Lysiphlebus ambiguus (Haliday): On *Eryngium planum*

Lysiphlebus fabarum (Marshall): *Amaranthus retroflexus, Anthemis sancti-johannis, Arctium* sp., *Arctium lappa, Beta vulgaris, Carduus crispus, Carduus rigens, Carduus* sp., *Callendula officinalis, Campanula rapunculoides, Chenopodium album, Chenopodium rubrum, Chenopodium bonus-henricus, Cirsium arvense, Cirsium palustre, Dahlia variabilis, Euonymus europaea, Melandrium album, Rumex crispus, Scorzonera parviflora, Torilis anthriscus, Urtica urens*

Praon abjectum (Haliday): On *Arctium* sp., *Cirsium rigens, Euonymus europaea, Philadelphus coronarius*

Trioxys acalephae (Marshall): On *Cirsium arvense, Arctium* sp.,

Trioxys angelicae (Haliday): On *Arctium* sp., *Beta vulgaris, Campanula rapunculoides, Chenopodium* sp., *Euonymus europaea, Philadelphus coronarius, Scorzonera parviflora*

   Aphis farinosa Gmel.

Ephedrus plagiator (Nees): On *Salix* sp.

Lysiphlebus ambiguus (Haliday): On *Salix* sp., *Salix repens* var. *rosmarinifolia, Salix viminalis*

Praon abjectum (Haliday): On *Salix* sp.

Trioxys acalephae (Marshall): On *Salix* sp.

Trioxys angelicae (Haliday): On *Salix* sp.

   Aphis galii-scabri (Schrk.)

Trioxys glaber Starý: On *Asperula cynanchica*

   Aphis genistae (Scop.)

Trioxys genistae Mackauer: On *Genista* sp.

   Aphis idaei v. d. G.

Ephedrus persicae Froggatt: On *Rubus idaeus*

Ephedrus plagiator (Nees): On *Rubus idaeus*

   Aphis intybi (Koch)

Lipolexis gracilis Förster: On *Cichorium intybus*

Lysiphlebus fabarum (Marshall): On *Cichorium intybus*

   Aphis lambersi (Börner)

Aphidius salicis Haliday: On *Daucus carota*

Lysiphlebus fabarum (Marshall): On *Daucus carota*

   Aphis mordwilkiana (Dobrowlj.)

Trioxys acalephae (Marshall): On *Rubus* sp.

   Aphis nasturtii (Kalt.)

Ephedrus plagiator (Nees): On *Rhamnus cathartica*

   Aphis newtoni Theo.

Lipolexis gracilis Förster: On *Iris variegata*

Lysiphlebus fabarum (Marshall): On *Iris variegata*

Aphis origani (Pass.)
Lipolexis gracilis Förster: On *Origanum vulgare*
   Aphis plantaginis (Goetze)
Lipolexis gracilis Förster: On *Plantago* sp.
Lysiphlebus fabarum (Marshall): On *Plantago* sp., *Plantago media, Plantago major*
   Aphis podagrariae Schrk.
Lysiphlebus fabarum (Marshall): On *Aegopodium podagraria*
Trioxys macroceratus Mackauer: On *Aegopodium podagraria*
   Aphis polygonata (Nevs.)
Lipolexis gracilis Förster: On *Polygonum aviculare*
Lysiphlebus fabarum (Marshall): On *Polygonum convolvulus*
   Aphis pomi (Deg.)
Ephedrus plagiator (Nees): On *Crataegus monogyna*
Lysiphlebus fabarum (Marshall): On *Malus silvestris, Pirus communis*
Trioxys angelicae (Haliday): On *Crataegus monogyna, Malus silvestris*
   Aphis poterii Börner
Lysiphlebus fabarum (Marshall): On *Sanguisorba minor*
   Aphis roepkei HRL
Lysiphlebus fabarum (Marshall): On *Potentilla reptans*
Paralipsis enervis (Nees): On *Potentilla anserina*
   Aphis rumicis L.
Lysiphlebus fabarum (Marshall): On *Rumex* sp.
   Aphis salviae (Walk.)
Lipolexis gracilis Förster: On *Salvia nemorosa, Salvia pratensis, Salvia* sp.
Lysiphlebus fabarum (Marshall): On *Salvia pratensis*
Trioxys acalephae (Marshall): On *Salvia nemorosa, Salvia pratensis*
Trioxys angelicae (Haliday): On *Salvia verticillata*
   Aphis sambuci L.
Praon abjectum (Haliday): On *Sambucus nigra*
Trioxys angelicae (Haliday): On *Sambucus nigra*
   Aphis schneideri (Börner)
Lysiphlebus ambiguus (Haliday): On *Ribes rubrum*
   Aphis spiraephaga Müller
Ephedrus plagiator (Nees): On *Spiraea* spp.
Trioxys acalephae (Marshall): On *Spiraea arguta*
Trioxys angelicae (Haliday): On *Spiraea arguta, Spiraea* spp.
   Aphis stachydis (Mordv.)
Lysiphlebus fabarum (Marshall): On *Stachys recta*
   Aphis taraxacicola (Börner)
Lipolexis gracilis Förster: On *Taraxacum officinale*
Lysiphlebus fabarum (Marshall): On *Taraxacum officinale*
   Aphis thomasi (Börner)

Lysiphlebus fabarum (Marshall): On *Scabiosa columbaria*
  Aphis urticata (F.)
Ephedrus plagiator (Nees): On *Urtica urens*
Lysiphlebus fabarum (Marshall): On *Urtica dioica*
Trioxys acalephae (Marshall): On *Urtica dioica*
  Aphis vandergooti Börner
Lysiphlebus fabarum (Marshall): On *Achillea millefolium*
  Aphis verbasci (Schrk.)
Lysiphlebus fabarum (Marshall): On *Verbascum austriacum, Verbascum* sp.
  Aphis viburni Scop.
Praon abjectum (Haliday): On *Viburnum opulus*
Trioxys angelicae (Haliday): On *Viburnum opulus*
  Aphis spp.
Ephedrus plagiator (Nees): On *Robinia pseudoacacia*
Lipolexis gracilis Förster: On *Peucedanum alsaticum, Rumex* sp., *Cirsium* sp.
Lysiphlebus fabarum (Marshall): On *Achillea ptarmica, Anthemis maczygia, Arnica sacchaliensis, Caragana arborescens, Cirsium arvense, Cirsium obium* sp., *Epilobium montanum, Eryngium campestre, Euphorbia* sp., *Galeopsis speciosa, Impatiens parviflora, Polygonum convolvulus, Rhamnus cathartica, Rumex* sp., *Torilis japonica, Trifolium pratense*
Praon abjectum (Haliday): On *Epilobium montanum, Epilobium parviflorum*
Trioxys acalephae (Marshall): On *Epilobium montanum, Rumex conglomeratus*
Trioxys angelicae (Haliday): On *Epilobium* sp., *Impatiens nolli-tangere, Malachium aquaticum, Tropaeolum majus*
  Aulacorthum chelidonii (Kalt.)
Ephedrus plagiator (Nees): On *Chelidonium majus*
  Aulacorthum dryopteridis Holman
Aphidius lonicerae Marshall: On *Dryopteris austriaca*
  Aulacorthum geranii (Kalt.)
Aphidius aulacorthi Starý: On *Erodium cicutarium, Geranium affine*
  Aulacorthum spp.
Aphidius aulacorthi Starý: On *Geranium robertianum, Naumburgia thyrsiflora, Potentilla argentea, Sanguisorba minor, Vincetoxicum officinale*
Ephedrus plagiator (Nees): On *Naumburgia thyrsiflora, Potentilla argentea*
  Brachycaudus ballotae Pass.
Paralipsis enervis (Nees): On *Ballota nigra*
  Brachycaudus cardui (L.)
Ephedrus plagiator (Nees): On *Prunus cerasifera, Prunus spinosa*
Lipolexis gracilis Förster: On *Carduus* sp., *Matricaria inodora*
Lysiphlebus fabarum (Marshall): On *Arctium lappa, Arctium* sp., *Carduus crispus, Carduus nutans, Carduus* sp., *Cirsium eriophorum, Leucanthemum vulgare, Matricaria inodora, Prunus spinosa*

194

Paralipsis enervis (Nees): On *Carduus crispus*
  Brachycaudus helichrysi (Kalt.)
Diaeretiella rapae (M'Int.): On *Senecio vulgaris*
Ephedrus persicae Froggatt: On *Anthemis* sp., *Hieracium laevigatum*, *Matricaria suaveolens*, *Prunus persica*
Lipolexis gracilis Förster: On *Melandrium album*, *Prunus persica*
Praon volucre (Haliday): On *Melandrium album*
Trioxys angelicae (Haliday): On *Prunus persica*
  Brachycaudus lychnidis (L.)
Ephedrus persicae Froggatt: On *Melandrium* sp., *Silene cucubalus*
Lysiphlebus melandriicola Starý: On *Melandrium album*
Praon volucre (Haliday): On *Melandrium album*
  Brachycaudus mordwilkoi HRL.
Lipolexis gracilis Förster: On *Echium vulgare*
Paralipsis enervis (Nees): On *Echium* sp.
  Brachycaudus rumexicolens Patch
Diaeretiella rapae (M'Int.): On *Rumex acetosella*
Lysiphlebus fabarum (Marsh.): On *Rumex acetosella*
  Brachycaudus tragopogonis (Kalt.)
Lysiphlebus fabarum (Marsh.) On *Tragopogon pratense*, *Tragopogon* sp.
  Brachycaudus spp.
Aphidius ervi Haliday: On *Senecio* sp. (record?)
Diaeretiella rapae (M'Int.): On *Rumex acetosella*, *Matricaria* sp.
Ephedrus persicae Froggatt: On *Prunus domestica*, *Melandrium* sp., *Prunus spinosa*
Ephedrus plagiator (Nees): On *Prunus domestica*, *Prunus* sp.
Lipolexis gracilis Förster: On *Arctium lappa*, *Carduus* sp., *Cirsium vulgare*, *Prunus domestica*, *Tanacetum vulgare*
Lysiphlebus fabarum (Marshall): On *Carduus acanthoides*, *Carduus crispus*, *Matricaria* sp., *Prunus domestica*, *Prunus persica*, *Tanacetum vulgare*
Paralipsis enervis (Nees): On *Arctium lappa*
Trioxys angelicae (Haliday): On *Rumex acetosella*
  Brevicoryne brassicae (L.)
Diaeretiella rapae (M'Int.): On *Alliaria vulgaris*, *Brassica napus*, *Brassica oleracea* var., *Brassica oleracea* var. *botrytis*, *Brassicaceae*, *Raphanus raphanistrum*, *Sinapis arvensis*
Praon volucre (Haliday): On *Brassica oleracea* var. *capitata*
  Buchneria pectinatae (Nördl.)
Pauesia infulata (Haliday): On *Abies alba*
  Calaphis spp.
Aphidius sicarius Mackauer: On *Betula* sp., *Betula verrucosa*
  Cavariella spp.
Aphidius salicis Haliday: On *Angelica silvestris*, *Anthriscus silvestris*, *Daucus carota*,

Salix spp., *Selinum carvifolia*
Ephedrus minor Stelfox: On *Heracleum sphondylium*
Trioxys brevicornis (Haliday): On *Daucus carota*
  Ceruraphis eriophori (Walk.)
Ephedrus plagiator (Nees): On *Viburnum opulus*
Trioxys angelicae (Haliday): On *Viburnum lantana*
  Chaitophorus sp.
Ephedrus plagiator (Nees): On *Populus* sp.
Lysiphlebus salicaphis (Fitch): On *Populus* sp.
  Chromaphis juglandicola (Kalt.)
Trioxys pallidus (Haliday): On *Juglans regia*
  Cinara spp.
Metaphidius aterrimus (Fahringer): On *Pinus silvestris*
Pauesia abietis (Marshall): On *Larix europaea*
Pauesia laricis (Haliday): On *Picea excelsa, Pinus silvestris*
Pauesia piceaecollis (Starý): On *Picea excelsa*
Pauesia pini (Haliday): On *Larix europaea, Pinus silvestris*
Pauesia silvestris (Starý): On *Pinus silvestris*
  Coloradoa achilleae HRL.
Lysaphidus arvensis Starý: On *Achillea millefolium*
  Coloradoa tanacetina (d. Gu.): On *Tanacetum vulgare*
Lysaphidus arvensis Starý: On *Achillea millefolium*
  Cryptomyzus ribis (L.)
Aphidius ribis Haliday: On *Ribes nigra, Ribes rubrum*
  Cryptosiphum artemisiae Bckt.
Ephedrus nacheri Quilis: On *Artemisia vulgaris*
  Cupressobium juniperi (Deg.)
Pauesia cupressobii (Starý): On *Juniperus communis*
Pauesia juniperorum (Starý): On *Juniperus communis*
  Dactynotus aeneus HRL.
Aphidius funebris Mackauer: On *Carduus acanthoides*
Ephedrus campestris Starý: On *Carduus nutans*
Trioxys centaureae (Haliday): On *Carduus crispus, Carduus* sp.
  Dactynotus campanulae (Kalt.)
Aphidius funebris Mackauer: On *Campanula* sp.
Praon dorsale (Haliday): On *Campanula rotundifolia*
Trioxys centaureae (Haliday): On *Campanula* sp.
  Dactynotus cichorii (Koch)
Aphidius funebris Mackauer: On *Cichorium intybus, Crepis biennis, Centaurea cyanus*
Ephedrus campestris Starý: On *Cichorium intybus, Leontodon hispidus*
Praon dorsale (Haliday): On *Cichorium intybus, Crepis biennis, Lapsana communis*
Trioxys centaureae (Haliday): On *Cichorium intybus, Crepis biennis*

Dactynotus cirsii (L.)

Aphidius funebris Mackauer: On *Cirsium* sp.

Dactynotus jaceae (L.)

Aphidius funebris Mackauer: On *Centaurea jacea, Centaurea scabiosa, Centaurea stoebe*

Ephedrus campestris Starý: On *Centaurea scabiosa, Centaurea stoebe*

Praon dorsale (Haliday): On *Centaurea scabiosa, Centaurea stoebe*

Trioxys centaureae (Haliday): On *Centaurea scabiosa, Centaurea stoebe*

Dactynotus muralis (Bckt.)

Aphidius funebris Mackauer: On *Mycelis muralis*

Ephedrus campestris Starý: On *Mycelis muralis*

Trioxys centaureae (Haliday): On *Mycelis muralis*

Dactynotus obscurus (Koch)

Aphidius funebris Mackauer: On *Hieracium silvaticum*

Ephedrus campestris Starý: On *Hieracium* sp.

Trioxys centaureae (Haliday): On *Hieracium* sp.

Dactynotus linariae (Koch)

Pron dorsale (Haliday): On *Aster linosyris*

Dactynotus picridis (F.)

Aphidius funebris Mackauer: On *Picris hieracioides*

Ephedrus campestris Starý: On *Picris hieracioides*

Dactynotus sonchi (L.)

Aphidius funebris Mackauer: On *Sonchus oleraceus*

Dactynotus taraxaci (Kalt.)

Praon dorsale (Haliday): On *Taraxacum officinale*

Dactynotus spp.

Aphidius funebris Mackauer: On *Campanula sibirica, Campanula* sp., *Carduus crispus, Carduus glaucus, Centaurea* sp., *Cirsium* sp., *Crepis biennis, Hieracium* sp., *Lactuca quercina, Sonchus oleraceus*

Diaeretiella rapae (M'Int.): On *Crepis biennis*

Ephedrus campestris Starý: On *Carlina vulgaris, Crepis biennis, Leucanthemum vulgare*

Drepanosiphon platanoidis (Schrk.)

Dyscritulus planiceps (Marshall): On *Acer pseudoplatanus*

Monoctonus pseudoplatani (Marshall): On *Acer pseudoplatanus*

Trioxys cirsii (Curtis): On *Acer pseudoplatanus*

Dysaphis crataegi (Kalt.)

Paralipsis enervis (Nees): On *Daucus carota*

Dysaphis devecta (Walk.)

Ephedrus persicae Froggatt: On *Malus silvestris*

Ephedrus plagiator (Nees): On *Malus silvestris*

Trioxys angelicae (Haliday): On *Malus silvestris*

Dysaphis sorbi (Kalt.)

Ephedrus persicae Froggatt: On *Sorbus aucuparia*

Ephedrus plagiator (Nees): On *Sorbus aucuparia*

   Dysaphis spp.

Ephedrus persicae Froggatt: On *Crataegus oxyacantha, Malus silvestris, Sorbus torminalis, Pirus communis*

Ephedrus plagiator (Nees): On *Crataegus* sp., *Malus silvestris, Pirus communis, Sorbus torminalis*

Lysiphlebus fabarum (Marshall): On *Arctium lappa*

Praon volucre (Haliday): On *Malus silvestris*

Trioxys angelicae (Haliday): On *Crataegus oxyacantha*

   Eucallipterus tiliae (L.)

Praon flavinode (Haliday): On *Tilia cordata, Tilia europaea, Tilia tomentosa*

Trioxys pallidus (Haliday): On *Tilia* sp.

   Galiobium langei Börner

Aphidius matricariae Haliday: On *Galium verum*

   Hayhurstia atriplicis (L.)

Diaeretiella rapae (M'Int.): On *Atriplex* sp., *Chenopodium album, Chenopodium* sp.

Ephedrus nacheri Quilis: On *Chenopodium album, Chenopodium* sp.

   Hyadaphis bupleuri Börner

Trioxys brevicornis (Haliday): On *Bupleurum falcatum*

   Hyadaphis mellifera (Hottes)

Ephedrus persicae Froggatt: On *Lonicera xylosteum*

   Hyadaphis sp.

Trioxys brevicornis (Haliday): On *Conium maculatum*

   Hyalopterus pruni (Geoffr.)

Aphidius transcaspicus Telenga: On *Prunus domestica, Phragmites communis* (Introduced)

Ephedrus plagiator (Nees): On *Prunus domestica*

Praon volucre (Haliday): On *Phragmites communis, Prunus armeniaca, Prunus domestica, Prunus spinosa*

   Hydaphias hofmanni Börner

Aphidius matricariae Haliday: On *Galium verum*

   Hydaphias spp.

Aphidius matricariae Haliday: On *Galium mollugo*

Lysiphlebus ambiguus (Haliday): On *Galium mollugo*

Trioxys parauctus Starý: On *Galium verum*

   Hyperomyzus lactucae (L.)

Aphidius sonchi Marshall: On *Sonchus asper, Sonchus oleraceus*

Ephedrus plagiator (Nees): On *Ribes grossularia, Ribes nigra*

Lysiphlebus fabarum (Marshall): On *Sonchus oleraceus*

Praon volucre (Haliday): On *Sonchus oleraceus*

   Impatientinum balsamines (Kalt.)

Monoctonus nervosus (Haliday): On *Impatiens nolli-tangere*

   Kallistaphis betulicola (Kalt.)

Aphidius sicarius Mackauer: On *Betula* sp.

   Linosiphon asperulophagus Holman

Aphidius matricariae Haliday: On *Asperula odorata*

   Linosiphon galiophagus (Wimsh.)

Aphidius matricariae Haliday: On *Galium silvaticum*

   Liosomaphis berberidis (Kalt.)

Aphidius hortensis Marshall: On *Berberis vulgaris*

Ephedrus plagiator (Nees): On *Berberis vulgaris*

   Lipaphis erysimi (Kalt.)

Lysaphidus erysimi Starý: On *Erysimum erysimoides*

   Macrosiphoniella absinthii (L.)

Aphidius absinthii Marshall: On *Artemisia absinthium*

Ephedrus campestris Starý: On *Artemisia absinthium*

Praon absinthii (Bignell): On *Artemisia absinthium*

   Macrosiphoniella artemisiae (B. d. F.)

Aphidius absinthii Marshall: On *Artemisia vulgaris*

Trioxys centaureae (Haliday): On *Artemisia vulgaris*

   Macrosiphoniella kaufmanni Börner

Aphidius absinthii Marshall: On *Achillea pontica, Achillea millefolium*

   Macrosiphoniella millefolii (Deg.)

Aphidius absinthii Marshall: On *Achillea millefolium, Achillea nobilis*

Ephedrus campestris Starý: On *Achillea millefolium, Achillea nobilis*

Praon absinthii (Bignell): On *Achillea nobilis, Achillea millefolium*

Trioxys centaureae (Haliday): On *Achillea millefolium*

   Macrosiphoniella pulvera (Walk.)

Aphidius absinthii Marshall: On *Artemisia maritima*

   Macrosiphoniella stägeri HRL.

Aphidius absinthii Marshall: On *Centaurea stoebe*

   Macrosiphoniella tanacetaria (Kalt.)

Praon absinthii (Bignell): On *Tanacetum vulgare*

Trioxys centaureae (Haliday): On *Chrysanthemum leucanthemum*

   Macrosiphoniella xeranthemi Bosh.

Aphidius absinthii Marshall: On *Xeranthemum foetidum*

   Macrosiphoniella spp.

Aphidius absinthii Marshall: On *Achillea sudetica, Artemisia campestris, Artemisia vulgaris, Matricaria chamomilla, Leucanthemum vulgare, Xeranthemum foetidum,*

   Macrosiphum daphnidis Börner

Aphidius lonicerae Marshall: On *Daphne mezereum*

   Macrosiphum euphorbiae Theo.

Praon volucre (Haliday): On *Euphorbia cyparissias*

Macrosiphum funestum (Macch.)
Aphidius rubi Starý: On *Rubus* sp.
Aphidius rosae Haliday: On *Rubus* sp.
   Macrosiphum gei (Koch)
Aphidius lonicerae Marshall: On *Geum* sp.
   Macrosiphum prenanthidis Börner
Aphidius lonicerae Marshall: On *Prenanthes purpurea*
Ephedrus plagiator (Nees): On *Prenanthes purpurea*
   Macrosiphum rosae (L.)
Aphidius ervi Haliday: On *Rosa* sp.
Aphidius rosae Haliday: On *Rosa* sp., *Scabiosa columbaria, Dipsacus* sp.
Ephedrus lacertosus (Haliday): On *Rosa* sp.
Ephedrus plagiator (Nees): On *Rosa* sp.
Praon rosaecola Starý: On *Rosa* sp.
Praon volucre (Haliday): On *Rosa* sp.

   Macrosiphum stellariae Theo.
Aphidius lonicerae Marshall: On *Stellaria holostea*
   Maculolachnus submacula (Walk.)
Pauesia maculolachni (Starý): On *Rosa* sp.
   Megoura viciae (Bckt.)
Aphidius megourae Starý: On *Lathyrus pratensis*
Praon dorsale (Haliday): On *Lathyrus* sp., *Lathyrus silvester*
   Metopeurum fuscoviride Stroyan
Aphidius tanacetarius Mackauer: On *Tanacetum vulgare*
Lysiphlebus hirticornis Mackauer: On *Tanacetum vulgare*
   Microlophium evansi (Theo.)
Aphidius ervi Haliday: On *Urtica dioica*
Praon volucre (Haliday): On *Urtica dioica*
Trioxys centaureae (Haliday): On *Urtica dioica*
   Microsiphum nudum Holman
Lysiphlebus fabarum (Marshall): On *Achillea nobilis*
   Mirotarsus cyparissiae (Koch)
Aphidius mirotarsi Starý: On *Euphorbia cyparissias*
   Myzaphis rosarum (Kalt.)
Ephedrus minor Stelfox: On *Rosa* sp.
   Myzella galeopsidis (Kalt.)
Aphidius ribis Haliday: On *Galeopsis tetrahit*
   Myzocallis carpini (Koch.)
Praon flavinode (Haliday): On *Carpinus betulus*
Trioxys pallidus (Haliday): On *Carpinus betulus*
   Myzocallis coryli (Goetze)
Ephedrus plagiator (Nees): On *Corylus avellana*

200

Myzodes auctus (Walk.)
Aphidius picipes (Nees): On *Cerastium tomentosum*
   Myzodes ligustri (Mosl.)
Ephedrus persicae Froggatt: On *Ligustrum aviculare*
Monoctonus cerasi (Marshall): On *Ligustrum aviculare*
   Myzodes persicae (Sulz.)
Aphidius picipes (Nees): On *Papaver dubium, Solanum tuberosum, Urtica urens*
Diaeretiella rapae (M'Int.): On *Beta vulgaris, Solanum tuberosum*
Trioxys angelicae (Haliday): On *Papaver dubium*
   Myzodes sp.
Diaeretiella rapae (M'Int.): On *Cerastium tomentosum, Malachium aquaticum, Urtica urens*
   Myzus ajugae (Schout.)
Aphidius matricariae Haliday: On *Ajuga genuensis, Ajuga reptans*
   Myzus cerasi (F.)
Aphidius matricariae Haliday: On *Prunus cerasus*
Ephedrus cerasicola Starý: On *Prunus avium*
Ephedrus persicae Froggatt: On *Prunus avium*
Ephedrus plagiator (Nees): On *Prunus avium, Prunus cerasus*
Lipolexis gracilis Förster: On *Prunus avium, Prunus cerasus*
   Nasonovia nigra HRL.
Aphidius hieraciorum Starý: On *Hieracium silvaticum*
Monoctonus angustivalvus Starý: On *Hieracium silvaticum*
Monoctonus crepidis (Haliday): On *Hieracium silvaticum, Hieracium* sp.
Praon pubescens Starý: On *Hieracium silvaticum*
   Nasonovia pilosellae (Börner)
Aphidius hieraciorum Starý: On *Hieracium pilosella*
Monoctonus crepidis (Haliday): On *Hieracium pilosella*
   Nasonovia ribisnigri (Mosl.)
Aphidius hieraciorum Starý: On *Hieracium* sp.
Monoctonus crepidis (Haliday): On *Hieracium bauhinii, Hieracium fallax, Hieracium pilosella, Hieracium pratense, Lapsana* sp.
Praon pubescens Starý: On *Hieracium* sp.
   Nasonovia spp.
Aphidius hieraciorum Starý: On *Hieracium* sp., *Hieracium echioides, Hieracium silvaticum, Hieracium pilosella*
Monoctonus crepidis (Haliday): On *Hieracium* sp., *Hieracium junceum, Hieracium echioides*
   Paczoskia major Börner
Aphidius funebris Mackauer: On *Echinops sphaerocephalus*
Lysiphlebus fabarum (Marshall): On *Echinops sphaerocephalus*
Praon dorsale (Haliday): On *Echinops sphaerocephalus*

Passerinia tetrarhoda (Walk.)

Ephedrus minor Stelfox: On *Rosa* sp.

Pemphigus sp.

Lysiphlebus fabarum (Marshall): On *Helichrysum arenarium*

Periphyllus villosus (Htg.)

Aphidius setiger Mackauer: On *Acer platanoides*

Trioxys falcatus Mackauer: On *Acer campestre, Acer platanoides*

Phalangomyzus oblongus (Mordv.)

Aphidius phalangomyzi Starý: On *Artemisia vulgaris*

Phorodon humuli (Schrk.)

Ephedrus persicae Froggatt: On *Humulus lupulus, Prunus domestica*

Ephedrus plagiator (Nees): On *Prunus domestica*

Trioxys humuli Mackauer: On *Humulus lupulus*

Prociphilus fraxini (Htg.)

Ephedrus plagiator (Nees): On *Fraxinus excelsior*

Protaphis carlinae (Börner)

Lysiphlebus fabarum (Marshall): On *Carlina* sp., *Carlina vulgaris*

Protolachnus agilis (Kalt.)

Diaeretus leucopterus (Haliday): On *Pinus* sp., *Pinus silvestris, Pinus nigra*

Praon bicolor Mackauer: On *Pinus silvestris*

Phyllaphis fagi (L.)

Trioxys phyllaphidis Mackauer: On *Fagus silvatica*

Pseudobrevicoryne erysimi Holman

Lysaphidus erysimi Starý: On *Erysimum crepidifolium, Erysimum dubium*

Pterocomma pilosum Bckt.

Aphidius pterocommae Ashmead: On *Salix caprea*

Pterocomma salicis (L.)

Aphidius pterocommae Ashmead: On *Salix amygdalina*

Pterocomma spp.

Aphidius pterocommae Ashmead: On *Populus* sp., *Salix* sp., *Salix caprea, Salix cinerea*

Rhopalosiphoninus sp.

Ephedrus lacertosus (Haliday): On *Oxalis acetosella*

Rhopalosiphum nymphaeae (L.)

Lysiphlebus fabarum (Marshall): On *Ranunculus* sp.

Praon necans Mackauer: On *Hydrochaeris morsus-ranae, Menianthes trifoliata, Nymphaea* sp.

Rhopalosiphum padi (L.)

Ephedrus persicae Froggatt: On *Padus racemosa*

Ephedrus plagiator (Nees): On *Padus racemosa, Secale cereale*

Praon abjectum (Haliday): On *Padus racemosa*

Trioxys angelicae (Haliday): On *Padus racemosa*

202

Rhopalosiphum sp.
Praon abjectum (Haliday): On *Amygdalus nana*
   Roepkea marchali (Börner)
Ephedrus persicae Froggatt: On *Prunus mahaleb*
   Schizaphis scirpi (Kittel)
Diaeretiella rapae (M'Intosh): On *Typha angustifolia*
Ephedrus plagiator (Nees): On *Typha angustifolia*
   Schizolachnus pineti (F.)
Pauesia unilachni (Gahan): On *Pinus nigra, Pinus silvestris*
   Schizolachnus sp.
Pauesia unilachni (Gahan): On *Pinus uliginosa*
   Schizoneura ulmi (L.)
Areopraon lepelleyi (Waterston): on *Ulmus campestris*
Ephedrus plagiator (Nees): On *Ulmus* sp.
   Semiaphis dauci (F.)
Aphidius salicis Haliday: On *Daucus carota*
   Sipha maydis (Pass.)
Lysiphlebus arvicola Starý: On *Medicago falcata*
   Sipha spp.
Lysiphlebus arvicola Starý: On *Agropyrum repens, Agropyrum* sp.
   Sitobium avenae (Fabr.)
Aphidius avenae Haliday: On *Festuca nemoralis, Hordeum distichum*
Aphidius pascuorum Marshall: On *Festuca nemoralis*
Ephedrus plagiator (Nees): On *Secale cereale*
Lysiphlebus fabarum (Marshall): On *Festuca* sp.
   Sitobium equiseti Holman
Aphidius equiseticola Starý: On *Equisetum silvaticum*
Ephedrus plagiator (Nees): On *Equisetum silvaticum*
Monoctonus caricis (Haliday): On *Equisetum silvaticum*
   Sitobium spp.
Aphidius avenae Haliday: On *Avena sativa, Hordeum distichum, Secale cereale*
Aphidius pascuorum Marshall: On *Grasses*
Diaeretiella rapae (M'Intosh): On *Lolium* sp.
Ephedrus plagiator (Nees): On *Avena sativa, Dactylis glomerata, Hordeum distichum, Rosa* sp., *Secale cereale, Triticum vulgare*
   Stomaphis sp.
Protaphidius wissmannii (Ratzeburg): On *Quercus* sp.
   Symydobius oblongus (v. Heyd.)
Trioxys betulae (Marshall): On *Betula* sp.
   Staegeriella necopinata (Börner)
Trioxys brevicornis (Haliday): On *Galium verum*
   Thelaxes dryophila (Schrk.)

Lysiphlebus thelaxis Starý: On *Quercus* sp.
  Therioaphis sp.
Praon exoletum (Nees): On *Melilotus albus, Medicago sativa*
  Tinocallis platani (Kalt.)
Praon flavinode (Haliday): On *Ulmus* sp.
Trioxys hortorum Starý: On *Ulmus effusa*
  Titanosiphon artemisiae (Koch)
Praon absinthii (Bignell): On *Artemisia campestris*
Trioxys pannonicus Starý: On *Artemisia campestris*
  Todolachnus abieticola (Chol.)
Pauesia grossa (Fahringer): On *Abies alba*
  Tuberculoides annulatus (Htg.)
Praon flavinode (Haliday): On *Quercus* sp.
Trioxys pallidus (Haliday): On *Quercus* spp.

204

XIV.

Natural limitation and control of aphids by parasites

The economic importance of aphid parasites in the control of pest aphids is immense. It is difficult to make an economic evaluation, being possibly estimated only in the countries with well-developed agriculture, but even there the effectiveness of the parasites can be ascertained only where the aphids on cultures are concerned, while their control of aphids on the other vegetation cannot be estimated. It is sufficient to say that according to American authors (Schlinger, 1960) mere importation and introduction of three species of parasites, conservation of native natural enemies, and use of effective insecticides reduced the $ 13 million problem of 1955 to about $ 2 million in 1958.

Damage of plants caused by aphids and possibilities of its prevention through aphid parasites

The aphids damage the plants roughly in the following ways:

1. Loss of plant juices by the sucking.

2. The reaction of plant tissue to the stimulation caused by aphid saliva is various: change of colour, curling of leaves, dwarfing of stem and leaves, galls, etc. The attacked trees grow weak, throw off their blossom and fruit, their wood does not ripen, and that is the reason they get easily frozen and often die after a longer or shorter time.

3. The liquid, viscous honey dew, containing much sugar and sprinkled around by the aphids, is very harmful. It covers the leaves with a coat that is soon congealed and makes assimilation and respiration very difficult. Besides, honey dew makes a medium for some saprophytic moulds.

In all three cases the aphidiids can have an effect upon the degree of damage to the plant depending on their effectiveness (see natural limitation, biological control, etc.).

4. Transmission of virus diseases. Considering that the rate of parasite development takes several days, parasitization does not mean an immediate killing of the aphid and prevention of the possibility of virus transmission. However, the parasites attack the aphids mostly in the lower instars and destroy them before reaching maturity and thus also the ability of migration. It means that on a certain plant affected by a virus disease, the parasitized aphids stay and die being unable to serve as virus vectors, be it by stylet-borne or circulative rout. Only in comparatively few cases, when the parasites infest the aphids in higher instars, the aphids reach maturity and are able to migrate; then the parasitization cannot prevent the virus transmission to another plant.

The parasitization of aphids on cultures then cannot prevent a possible infection of the plant by virus diseases through the sucking of the aphids. However, the parasitization limits the possibility of the spread of the infection to other specimens of the same species of host plant. On the other hand, by killing the aphids on economically indifferent plants the parasites decrease their number and, indirectly, the possibility of transmission and spread of the virus diseases on cultures.

Examples. Monoecious aphids. *Aphis idaei* v. d. G. is a vector of many viroses on *Rubus*. The parasites cannot prevent the spread of the aphid to other plants, but by decreasing the number of aphids on the infected plant they limit the spread and transmission of the disease to other plants. A similar example is performed by *Brevicoryne brassicae* (L.) on cabbage.

Dioecious aphids. *Sitobium avenae* (F.) is attacked on its primary host plant *(Rosa)* by parasites that control it, and a smaller number of aphids can migrate to the secondary host plants, on corn. There other parasitic species control the aphid again, so that on one hand a smaller number of aphids can migrate to other corn plants and transmit viruses, on the other, a small number can migrate in autumn from corn to the primary host plants (source of the pest aphid occurrence in the next year).

Natural limitation of aphids by parasites

The natural limitation of aphids by parasites is rather changeable and depends on a number of various factors. Generally it is true, that the effectiveness of the parasites can be better ascertained and presumed in long-lasting communities, i.e. particularly in forest type habitats and in perennial crops. In the perennial crops, where the number of parasites greatly depends on the environment, their effectiveness is difficult to forecast. From the viewpoint of the importance of the natural limitation just the long-lasting communities are significant, because the effectiveness of the parasites can be reckoned with, while in the perennial crops it is never guaranteed.

The parasite's effectiveness depends roughly on the following factors:

1. Host and parasite population densities. They are most suitable when the

percentage of parasitization is the highest, but the percentage of superparasitization remains low.

2. Host and parasite fecundity. Aphids are a group characterized by a numerous progeny produced in a very short period.

It is important that the parasites infest the aphids of lower instars and kill them before their reaching maturity, so that the parasitization means death of the aphid without producing any progeny.

The aphids develop in spring at a lower average temperature than their parasites. That is the reason the fundatrices with great reproductive capacity are not, or rarely, infested in spring, and only the second generation of fundatrigeniae, usually numerous and producing a numerous progeny, is attacked by the parasites. Thus the dioecious aphid species on the primary host plants become too numerous before the parasites are able to reach an effective population density. Therefore the parasites are not usually able to control the aphids on the primary host plants before their migration to the secondary ones. Similarly on cultural crops where the aphids are at first spread in foci, the aphid colonies develop and reach a large quantity before the parasites can spread successfully all over the field from the environmental habitats including foci of the parasites. From this viewpoint the situation in the monoecious aphids, where a certain relation between the host and parasite populations may develop during the season, is more advantageous.

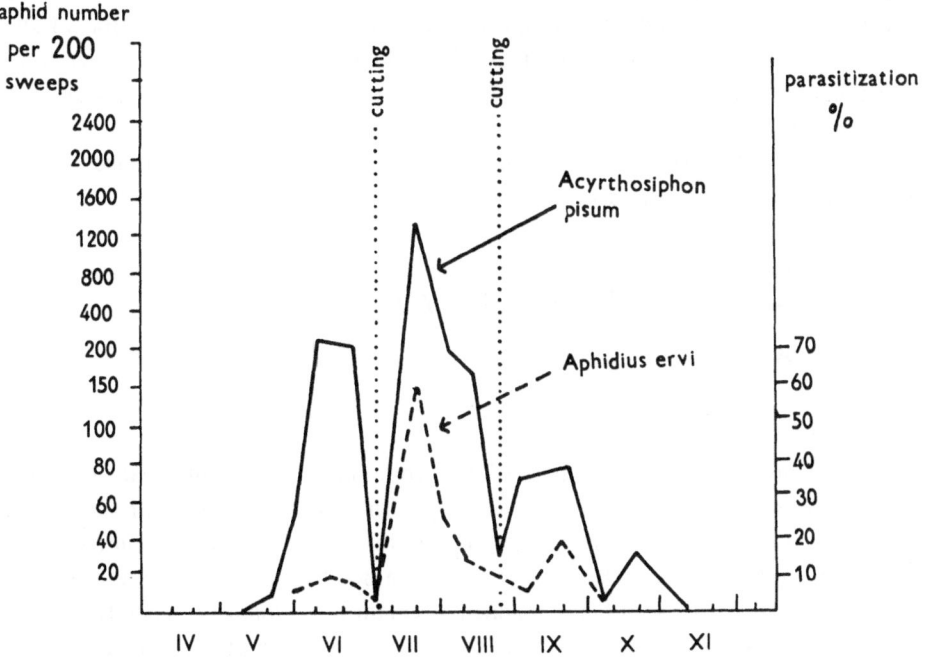

Fig. 8. Natural limitation of *Acyrthosiphon pisum* Harris by *Aphidius ervi* Hal. Western Bohemia, red clover field, 1956.

3. Presence of certain host instars. The parasites prefer certain host instars, and that is why they do not attack alate females migrating to a plant, but their progeny of II—III instar. Therefore the parasites are unable to prevent reproduction of the newly spread aphids on cultural plants.

4. Host and parasite behaviour is very specific, in many cases it is more important than the phylogenetic relationship of the host. This is in the connection with factors influencing the food specificity of the parasites (see chapter VII).

5. Food sources of adult parasites. The aphidiids probably feed on honey dew in the field, so that sources of food of the adult parasites may be found at the very place as the host. It is very advantageous when introducing a parasitic species in a new environment.

6. The opportunity of both sexes to meet. Little has been known about the olfactory orientation of the parasites, but it seems that the orientation of both sexes of adult parasites to the host is most important, as there both sexes are supposed to meet.

7. The presence of foci of parasites in the neighbourhood is one of the most important factors affecting the effectiveness of the aphid parasites in the open, particularly on perennial crops (see chapter XI).

8. Natural enemies. Secondary parasites sometimes influence in a high degree the effectiveness of the primary ones, particularly in the period of their maximal effect that follows after the maximum of the aphid outbreak. E.g. in Czechoslovakia the influence of the secondary parasites on the effectiveness of *Aphidius ervi* Hal., parasite of the pea aphid *Acyrthosiphon pisum* (Harris) on red-clover and alfalfa, was studied and can be mentioned here. All predators of aphids can be mentioned as facultative natural enemies of *Aphidiidae*, but their influence does not seem to be very important.

9. Abiotic factors (temperature, humidity, light) affect the occurrence both of aphids and parasites in the open in a remarkable way.

Control of aphids by parasites

Biological control

Literary review

Experiments on the employment of the aphidiids in the biological control of aphids were started more than fifty years ago. In 1907 the workers of the Kansas University used *Lysiphlebus testaceipes* (Cresson) in the biological control of the aphid *Schizaphis graminum* Rond. It was ascertained that the aphidiid appears only at the time when the aphid has reached a calamitous quantity. Therefore plants with attacked aphids were picked in southern areas and planted in middle and northern Kansas. Several hundred million aphidiids were thus artificially distributed all over Kansas. However, the work was done without the experimental basis and

208

the effect was estimated according to information from farmers, who often gave in to enthusiasm and reported the decrease of aphids in the fields at a time when the parasites could have not yet reproduced. Nevertheless, that historical experiment proved that the parasite can perform a great effect under suitable conditions.

Similar experiments were made by Hazelhoff (1929), transferring the native parasites from old to young fields to control the sugar cane aphid.

The introduction of *Lysiphlebus testaceipes* (Cress.) in Hawaii was successful (Williams, 1936).

Observations on the artificially introduced parasites of aphids infesting wheat were made in England (Arthur, 1945).

A really extensive action propagating the aphidiids in the biological control has been performed comparatively recently by the workers of the University of California at Riverside. The first problem of the research was the control of *Therioaphis trifolii* pesting on alfalfa. The aphid had been brought in America, but without parasites. Therefore research was made in Europe, Africa, and Asia and the parasites were introduced in the Riverside laboratories. Three hymenopterous parasites were obtained, two of them belonging to the family *Aphidiidae* − *Trioxys complanatus* Quilis (= *utilis* Muesebeck) and *Praon exoletum* (Nees) (= *palitans* Muesebeck). All those species were reared indoors and liberated in Californian fields, became successfully established, and have been playing an increasingly important role in the biological control of *Therioaphis trifolii* (Mon.). Gradually these effective parasites were introduced in other areas in America.

Another problem was the introduction of *Aphidius ervi* Hal. (= *smithi* Sharma and Subba Rao) from India against the aphid *Acyrthosiphon pisum* (Harris), pesting also on alfalfa. The introduction was successful, too, and the parasite has been exerting a dominant role among the natural enemies in the biological control of the pea aphid in California.

Similarly, the introduction of *Trioxys pallidus* (Hal.) against *Chromaphis juglandicola* (Kalt.) that has long been a major accidentally introduced pest of walnut in California, was also successful. The aphid is attacked by certain native parasites in California, but the natural limitation had been low and therefore *Trioxys pallidus* (Hal.) was introduced. Now the parasite is widespread in South California, and will probably control the walnut aphid successfully.

Biological control in Czechoslovakia

Experiments with the introduction of three aphidiid species have been made in Czechoslovakia till now. When introducing the aphidiids we started from the harmfulness of the aphid species against which the parasite was to be employed, and from the effectiveness of the native parasites. Then, regarding our studies of the foci and host specificity of the parasites, the species infesting aphids (esp. monoecious) living in long-lasting, at least 1 year, communities were selected. In the parasites of the dioecious aphid species like *Hyalopterus pruni* (Geoffr.) the problem to what type of habitat the aphid migrates, and the question of alternative hosts, etc., were solved first.

Aphidius "smithi Subba Rao and Sharma" was introduced in Czechoslovakia from laboratory breedings of the California University at Riverside, where it had been introduced from India as an effective parasite of the pea aphid, monoecious species pesting on forage crops in Czechoslovakia, too. The native parasites of the aphid in Czechoslovakia, *Aphidius ervi* Haliday and *Praon dorsale* (Hal.), are effective: according to our observations the effectiveness of *Aphidius ervi* Hal. was almost 60% in 1956 and 90% in 1960, but only in the period when the plants were heavily damaged due to the aphid outbreak. That was the reason we tried to introduce such parasitic species which could infest the aphids early in spring and prevent their outbreak.

Aphidius "smithi" was reared in the laboratory and the most important ecological data were obtained. Nevertheless, after the laboratory experiments the problem was left, as we found by comparison with the material collected in Central Asia and southern Europe that *Aphidius "smithi"* does not represent a valid species but only a colour variety of the widely distributed *Aphidius ervi* Haliday, which is also the native parasite of the pea aphid in Czechoslovakia.

Aphidius megourae Starý was introduced in Czechoslovakia from the European part of USSR against *Megoura viciae* Buckt., which is a facultative pest of some forage crops in Czechoslovakia. As bean was in the past, and probably will be again in the near future an important forage crop in Czechoslovakia, the study of the parasites of its aphid pests is of a certain preventive importance, though first *Aphidius megourae* Starý has to serve as a model object to show the scheme of the laboratory and field work on an aphidiid species that is to be introduced in a certain country.

The parasite was found to be rather effective, as the average number of parasitized aphids is about 115 per one female parasite at $18-24°C$ of the laboratory temperature. Nevertheless, *Aphidius megourae* Starý is relatively not a very important parasite and is probably specialized on *Megoura viciae* Bekt. which has not been a serious aphid pest in Czechoslovakia till now. Therefore experiments have been made on the propagation of *Acyrthosiphon pisum* Harris, a serious aphid pest on forage crops in Czechoslovakia, employed as an unnatural host of *Aphidius megourae* Bckt. The parasite was initially established in Czechoslovakia in southern Bohemia in 1963 (see Starý, 1964c).

Aphidius transcaspicus Telenga is distributed as a parasite of *Hyalopterus pruni* (Geoffr.), a serious aphid pest in orchards, from southern Europe to Central Asia. In Czechoslovakia *Hyalopterus pruni* (Geoffr.) is a serious pest in orchards on *Prunus domestica, Prunus persica* and *Prunus armeniaca, Praon volucre* (Hal.) and *Ephedrus plagiator* (Nees), widely polyphagous, being its only native parasites. As *Hyalopterus pruni* (Geoffr.) occurs both on primary and secondary host plants in habitats of the $\pm$ same type, it is probable that the aphid will be infested by the parasites throughout the season, so that the problem of alternative hosts seems to be insignificant in this case (compare with Starý, 1965).

Rearing and mass production

Mass rearing. For the laboratory breedings both of aphids and parasites, silon cages were found to be most suitable. The breeding method, or the choice of aphids and parasites depends on other technical possibilities. Many aphid species can be bred under fluorescent light on young host plants (bean, corn, potato, pea, etc.). In such a case, especially if we have to deal with dioecious aphids, it is necessary to regulate the period of daylight to make the aphids produce continually a parthe – nogenetic progeny only (18 hours) and thus to prevent the break of breeding due to the existence of sexuales and eggs. Cages with aphids must be sprinkled. The growing of plants in glass pots, which can be replaced by new ones, is most suitable. In such

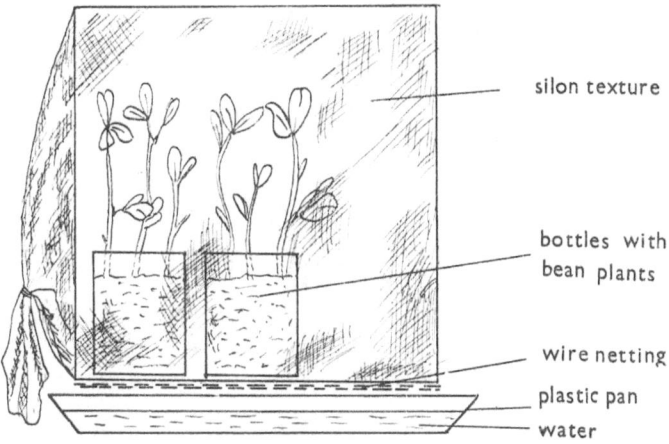

silon texture

bottles with
bean plants

wire netting
plastic pan
water

Fig. 9. Rearing silon cage for breeding aphids and parasites.

a case it is useful to have a few pots with plants of the same age in the cages, because older plants gradually fade and the aphids move on to younger ones. Always it is advantageous to cut off the tops with aphid colonies and leave them for 1 – 2 days in the cage to let the aphids move to other plants.

The parasites are reared under the same conditions as the aphids. They only require more frequent dewing, and for the longevity of the adults it is recommendable to add a few drops of honey or agar-honey in the cage. The cut plants must be left in the cage for a longer time (because the mummified aphids are on them), until the emergence of the parasites. The aphids must be supplied regularly to prevent the lack of them and dying out of the breeding, particularly when effective parasitic species are concerned, and to prevent too high a percentage of superparasitism.

The breeding of aphids and parasites is easier in a glasshouse. There it is necessary to regulate the time of illumination so that the length of daylight is constant and the cycle of the aphids does not change. The other principles of breeding are the same as in the laboratory. In the cages, if they are used in the glasshouse, the plants should be planted in strips with certain breaks and the growth renewed all the time, because the overpopulation of aphids destroy it. If there is an unbroken growth of host plants, the same procedure should be kept. The aphids and parasites in the glasshouse should be liberated in the growth in foci, or isolators can be used until a sufficient population density is accomplished. The capturing of parasites is easiest on the ceiling windows of the glasshouse, especially when in summer the wall windows are painted white. In the hot summer period the airing of the glasshouse is necessary, and some windows can be replaced by a silon texture.

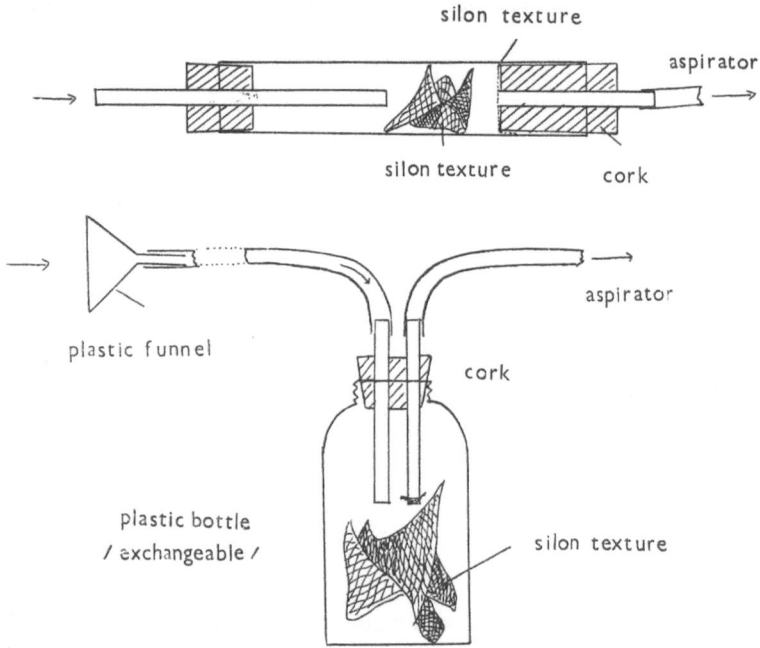

Fig. 10. Apparatus for parasites mass-collection.

Mass collection. For catching the adult parasites in the laboratory breedings the commonly used suction collector is most useful. A minor adjustment of the collector for the mass breedings is advisable, so that the parasites are sucked right in plastic bottles (Plate XXII, Fig. 49) in which they can be transported. It is necessary to put a piece of silon texture in the collector to prevent the parasites from mechanical injury which might be caused if they fell quickly in the collector. Also it is advantageous to use such an aspirator where the intensity of sucking can be controlled (Plate XXII, Fig. 48).

Storage and transport. The storage and transport of the parasites are safest in plastic bottles covered with silon (with a piece of the silon texture inside) or in

212

smaller silon isolators (Fig. 9). In both cases they are then preserved in cold rooms (about + 10°C), as may be required. The occasional sprinkling and feeding with honey is advisable. If the infested aphids are liberated together with the parasites, they must be transported separately, best with leaves of the plants between the pieces of silon texture, in small silon cages.

Initial establishment

On the conditions of a successful introduction of the parasites is their good initial establishment, which has two phases:

Selection of localities. In our opinion it is advantageous to liberate the introduced parasite on the one hand in a more or less original kind of habitat, on the other hand in a new habitatat to succeed in initial establishment. Both depend on the ecological characteristics of the species. As results from chapter VII., the parasites are strongly fixed on a certain type of habitat; this should be observed at the introduction especially of the parasites of the dioecious aphids which change the type of habitat during their migration. E.g. the initial establishment of a parasite which is typical for forest type habitat, is not very probable to be successful in steppe type habitats (see the case of *Trioxys complanatus* Quilis and *Trioxys pallidus* Hal.).

Liberation technique. The parasites and parasitized aphids are liberated at chosen places in favourable weather. The isolation of the adult parasites on a plant attacked by aphids under a silon sack for a few hours is most suitable for the initial establishment, as a large number of aphids is parasitized at one place and a focus of parasites arises. The living parasitized or mummified aphids are liberated in a similar manner; their occurrence in the place of the initial establishment ensures a longer-lasting occurrence of the parasite. It is advisable to liberate adults 1 − 2 days old, because such females have strong oviposition stimuli and attack the host immediately at the place of liberation, so that the formation of focus is more probable. The liberation should not be made only once, but repeated several times at the same place. If it is possible, a big isolator placed in the field is advantageous, as the liberated parasites concentrate at one place forming a large focus, and a successful establishment is ensured.

A scheme of biological control program

1. The investigation of native parasites, their life history, host specificity and effectiveness.

2. The investigation of parasites of the host in the general distribution area and of the basic data of their ecology.

3. Selection of suitable species for introduction.

4. Preparation of the laboratory breedings of the host and introduced parasite.

5. Obtaining of the living material of the introduced parasite.

6. Laboratory breedings. Investigation in the basic ecological data of the introduced parasite.

7. Mass-rearing.

8. Choice of localities for the initial establishment, classification of the environment.

9. Initial establishment and further research on introduction.

10. The parasite's effectiveness in the field, inter-specific competition among the native and introduced parasites, secondary parasites, etc.

11. Food specificity (alternative hosts, etc.) of the introduced species and its requirements on the new environment.

12. Integrated control program, etc.

Integrated control

Theory

The one-sided application of the chemical or biological control of insects showed that each type of control had both positive and negative effects. In the chemical control, particularly in the past, the insecticides were applied without thinking of the future and balance of the treated community, and the immediate control of the pest's outbreak was the only concern. New problems have arisen with the application of insecticides:

1. Many arthropod species became resistant to certain insecticides, and, as a result, new and new kinds of insecticides had to be applied and the problem became increasingly complicated.

2. The chemical treatment often resulted in the disturbance of the community's balance, and, consequently, new pest species, economically indifferent before, began to overpopulate, as their natural enemies were killed by the chemical treatment.

3. Many arthropod species adapted themselves to the insecticides, so that more effective forms of treatment or new insecticides had to be applied. That brought complications in the agricultural practice, and new and new chemical treatment programs were required.

4. The insecticides leave toxic residues in food and forage. This is one of the most important negative aspects of the chemical control, demanding more attention at present.

5. Many insecticides are lethal for man, and manipulation with them is dangerous.

In the modern chemical control program new methods are worked out, using selective insecticides which are to kill only a certain pest species, while the natural enemies and all the community are not influenced.

214

Recently all negative as well as positive aspects of the individual types of biological control have been valued, and a new conception of integrated control, involving cultural, physical, biological and chemical types, has been proposed. The share of the individual constituents is very variable, depending on the object. The present aim of the integrated control is a program that would allow to keep or re-establish the natural balance in a community under an effective control.

In some cases the augmentation of natural enemies is a basic condition of the integrated control. It can occur roughly in four ways:

1. Introduction of parasites. Sometimes the effectiveness of the native parasites is very low, and the introduction of a more effective species from other countries is necessary. Generally it is the easiest way, and many examples give evidence of its success (see the biological control of *Therioaphis trifolii* (Mon.), *Chromaphis juglandicola*, etc., in USA).

2. Artificial inoculation of the host at the time of low density. This manner seems to be satisfactory in the aphidiids, particularly on perennial crops. Naturally, the peculiarities of the life histories of the host and parasites must be observed.

3. Modification of the environment is a long-termed provision, but apparently the most effective and long-lasting in perspective. Irrigation changes, development of a greater heterogeneity of plants, and introduction of a covercrop, strip treatment with chemicals, development of refuges, etc. are usually involved.

4. Selective breeding of parasites. It has not yet been accomplished in the aphidiids, but experiments on other groups show that the method is promising.

Considering that the integrated control of aphids is a comparatively new method, it has been recently applied against the aphids *Therioaphis trifolii* (Mon.) and *Chromaphis juglandicola* (Kalt.) in USA.

Integrated control in Czechoslovakia

The following problems were solved in the range of integrated control:

Selective insecticides. *Acyrthosiphon pisum* (Harris) and its parasite *Aphidius ervi* Hal. (Obrtel, 1961). In the year 1960 a calamitous outbreak of pea aphid occurred at many places in Czechoslovakia on forage crops. In that connection the 30−40% of parasitization by *Aphidius ervi* Hal. was recorded in the spring, i.e. the parasite was very effective in natural limitation of the aphid pest. Therefore the effect of demetion and malathion on the instars of *Aphidius ervi* Hal. was investigated under laboratory conditions. It was ascertained that the selective insecticides have a lethal influence on instars of *Aphidius ervi* Hal. inside the mummified aphids, if submitted to the effect of unchanging residues and evaporation in a shut space. In the field the insecticides have a far weaker and insignificant lethal effect on the larvae and pupae of *Aphidius ervi* Hal. inside the mummified aphids. The adult parasites are killed by the residues of the insecticides applied both in the laboratory and field. That is the

215

reason the insecticidal treatment was proposed to be made at a time when parasites would be least afflicted. The laboratory experiments showed that the mortality of adult parasites depends most on the persistance of the residues of insecticides and length of the pupal stage in *Aphidius ervi* Hal. Therefore the treatment should be timed so that its residual effect disappears within a week at the most after the occurrence of the first mummified aphids.

Aphis fabae Scop. and its parasites (Zelený, unpubl. suggestion). The effect of the systemic insecticides was examined, as the black bean aphid is one of the serious pests of sugar beet in Czechoslovakia and its chemical control had been widely applied. The effect of the insecticides on the main parasite of the aphid in steppe type habitats — *Lysiphlebus fabarum* (Marsh.) — was investigated.

Augmentation of aphid parasites. Development of uncultivated and untreated areas. A profound analysis of the surroundings of cultivated areas from the viewpoint of the aphids and their parasites has been accomplished in Czechoslovakia in connection with the investigation of the foci of aphid parasites (see chapter XI). It showed the importance of the problem: the adjacent areas often directly influence the composition and effectiveness of parasites in the cultivated areas. The study of the foci — refuges of the parasites aims for the formation of universal foci of beneficial insects at suitable places or habitats in agricultural areas where they could concentrate, from there spread to cultures and avoid the harmful effect of chemical treatment. Besides the problem discussed in chapter XI it must be emphasized that at the analysis of the fauna of aphids and parasites the peculiarities in their life history should be observed. For example in orchards various species of the genus *Dysaphis* are pesting on apple-trees, and *Myzus cerasi* (F.) on cherry-trees causing the curling of leaves, deformation of tops, etc. They are mostly attacked by *Ephedrus persicae* Froggatt, which is a specialized parasite of the leaf curling aphids. By the analysis of the environmental fauna we can ascertain that the alternative host of the parasite is the aphid *Dysaphis sorbi* (Kalt.), developing at the same time as the *Dysaphis* species pesting in orchards, but in contrast to them being an economically indifferent species. A superficial analysis may suggest that a rowan culture near the orchards ensures a sufficient number of *Dysaphis sorbi* (Kalt.) and, consequently, the augmentation of the parasites of *Dysaphis* spp. on *Malus*. However, a more detailed investigation of the life-history of *Ephedrus persicae* Froggatt showed that in central Europe the species is very adaptable to the life-cycle of the host, which migrates at the end of spring to the secondary host plants in steppe habitats. Adaptation of the parasite to the life-cycle of the host appears in the larval diapause performed by a variable percentage of the parasites at the end of spring and beginning of summer; the diapause lasts until the spring of the next year, when the parasites emerge and again attack *Dysaphis*-species on the primary host plants. The parasites infesting *Dysaphis sorbi* (Kalt.), the migration cycle of which resembles that in the other *Dysaphis*-species, enter the diapause, too. It results that the planting of *Sorbus* near the orchards is prevented by the peculiarities in its life-history.

216

In comparison with the above case the support of the parasites which do not enter the diapause, occurring throughout the vegetative season in the same type of habitat, is much more advantageous. For example *Trioxys angelicae* (Hal.), parasitizing many pest aphids in the forest type habitats in spring, particularly *Aphis fabae* Scop. on *Euonymus europaea,* is such a parasite as well as *Ephedrus plagiator* (Nees), attacking besides *Aphis fabae* Scop. also *Rhopalosiphum padi* (L.), etc. When the dioecious aphids like *Aphis fabae* Scop. leave the forest type habitats, *Trioxys angelicae* (Hal.) attacks other, especially monoecious aphids in the same habitat. Since *Trioxys angelicae* (Hal.) limits, at least partly, the outbreak of *Aphis fabae* Scop. on the primary host plants in spring before the aphid migrates to cultures (sugar beet), it should be supported in the forest type habitats through the plants with the indifferent aphidofauna like *Spiraea* (*Aphis spiraephaga* Müller), *Laburnum vulgare* (*Aphis laburni* Kalt.), etc.

The introduction of a covercrop and development of a greater heterogeneity of plants seems to be easiest to put in practice for a higher effectiveness of the parasites and parasitic insects in general. Therefore the analysis of the aphidofauna of honey plants of Czechoslovakia was made (Starý, 1962a). It is well-known that the honey plants are important for four reasons: 1. They are a source of pollen and nectar for pollinators, 2. The collecting of nectar is sometimes important economically (bee-keeping), 3. nectar and pollen are a source of food of many adult parasitic insects, 4. On the flowering honey plants both sexes of the parasitic insects meet.

Like most plants, the honey plants have a special aphidofauna, including pests and economically indifferent aphids. For the augmentation of parasites it is good if such plants are present the aphidofauna of which includes mostly or exclusively indifferent aphid species acting as alternative hosts of the aphidiids. The parasites partly killing the aphids in the habitat, partly spreading to the neighbouring habitats where they attack the pest aphids on cultures, are thus supported. In short, the indifferent aphidofauna includes a) alternative hosts of the effective aphidiid species, b) producers of honey dew, i.e. food of adults of many parasitic insects.

A list of honey plants, their aphidofauna and its aphid parasites.

In different species of honey plants only such aphid species are listed from which aphidiid parasites were bred in Czechoslovakia. For the division of honey plants as to the habitats the general scheme used in the literature on apiculture has been accepted. It is necessary to note that the different groups of habitats cannot be sharply separated from each other and this feature corresponds to the composition of the fauna, too.

Habitats of kinds 1 — 5 belong to the forest type, 6 — 7 to steppe type habitats.

1. Coniferous forests

Abies alba — *Buchneria pectinatae* (Nördl.): *Pauesia infulata* (Hal.), *Todolachnus abieticola* (Chol.): *Pauesia grossa* (Fahr.)

217

Juniperus communis — *Cupressobium juniperi* (Deg.): *Pauesia cupressobii* (Starý), *Pauesia juniperorum* (Starý)

Larix decidua — *Cinara* spp.: *Pauesia abietis* (Hal.)

Picea excelsa — *Cinara* spp.: *Pauesia laricis* (Hal.), *Pauesia piceaecollis* (Starý)

Pinus silvestris — *Cinara* spp.: *Metaphidius aterrimus* (Fahr.), *Pauesia laricis* (Hal.), *Pauesia pini* (Hal.), *Pauesia silvestris* (Starý), *Schizolachnus pineti* (F.): *Pauesia unilachni* (Gahan), *Protolachnus agilis* (Kalt.): *Diaeretus leucopterus* (Hal.), *Praon bicolor* Mack.

2. Mixed forests

Acer platanoides — *Drepanosiphon platanoidis* (Schrk.): *Trioxys cirsii* (Curt.), *Periphyllus villosus* (Htg.): *Trioxys falcatus* Mack.

Acer pseudoplatanus — *Drepanosiphon platanoidis* (Schrk.): *Dyscritulus planiceps* (Marsh.), *Monoctonus pseudoplatani* (Marsh.), *Trioxys cirsii* (Curt.)

Berberis vulgaris — *Liosomaphis berberidis* (Kalt.): *Ephedrus plagiator* (Nees), *Aphidius hortensis* (Marsh.)

Epilobium spp. — *Aphis* sp.: *Praon abjectum* (Hal.)

Euonymus europaea — *Aphis fabae* Scop.: *Ephedrus plagiator* (Nees), *Praon abjectum* (Hal.), *Trioxys angelicae* (Hal.)

Padus racemosa — *Rhopalosiphum padi* (L.): *Ephedrus plagiator* (Nees), *Ephedrus persicae* Frog., *Praon abjectum* (Hal.), *Trioxys angelicae* (Hal.)

Prunus spinosa — *Brachycaudus cardui* (L.): *Ephedrus plagiator* (Nees)

Quercus spp. — *Tuberculoides annulatus* (Htg.): *Praon flavinode* (Hal.), *Trioxys pallidus* (Hal.)

Ribes grossularia — *Hyperomyzus lactucae* (L.): *Ephedrus plagiator* (Nees)

Robinia pseudoacacia — *Aphis craccivora* (Kalt.): *Ephedrus plagiator* (Nees), *Praon abjectum* (Hal.), *Trioxys acalephae* (Marsh.)

Salix caprea — *Aphis farinosa* (Gmel.): *Praon abjectum* (Hal.), *Lysiphlebus ambiguus* (Hal.), *Trioxys angelicae* (Hal.), *Trioxys acalephae* (Marsh.), *Pterocomma salicis* (L.): *Aphidius pterocommae* Ashm., *Pterocomma pilosum* Bckt.: *Aphidius pterocommae* Ashm.

Sambucus nigra — *Aphis sambuci* L.: *Trioxys angelicae* (Hal.), *Praon objectum* (Hal.)

Sarothamnus scoparius — *Acyrthosiphon spartii* (Koch): *Ephedrus plagiator* (Nees)

Sorbus aucuparia — *Dysaphis sorbi* (Kalt.): *Ephedrus plagiator* (Nees), *Ephedrus persicae* Frog.

Tilia spp. — *Eucallipterus tiliae* (L.): *Praon flavinode* (Hal.), *Trioxys pallidus* (Hal.)

Ulmus carpinifolia — *Schizoneura ulmi* (L.): *Ephedrus plagiator* (Nees), *Areopraon lepelleyi* (Waterst.)

Viburnum opulus — *Aphis viburni* Scop.: *Trioxys angelicae* (Hal.), *Praon abjectum* (Hal.), *Ceruraphis eriophori* (Walk.): *Ephedrus plagiator* (Nees), *Trioxys angelicae* (Hal.)

3. Parks and gardens

Acer platanoides — *Drepanosiphon platanoidis* (Schrk.): *Trioxys cirsii* (Curt.), *Periphyllus villosus* (Htg.): *Trioxys falcatus* Mack.

Acer pseudoplatanus — *Drepanosiphon platanoidis* (Schrk.): *Dyscritulus planiceps* (Marsh.), *Monoctonus pseudoplatani* (Marsh.), *Trioxys cirsii* (Curt.)

Caragana arborescens — *Acyrthosiphon caraganae* (Chol.): *Aphidius caraganae* Starý, *Ephedrus plagiator* (Nees), *Praon volucre* (Hal.), *Toxares deltiger* (Hal.), *Trioxys angelicae* (Hal.), *Aphis craccivora* (Koch.): *Trioxys angelicae* (Hal.)

Carpinus betulus — *Myzocallis carpini* (Koch): *Praon flavinode* (Hal.)

Euonymus europaea — *Aphis cognatella* Jones: *Trioxys angelicae* (Hal.), *Aphis fabae* Scop.: *Ephedrus plagiator* (Nees), *Praon abjectum* (Hal.), *Trioxys angelicae* (Hal.)

Fraxinus excelsior — *Prociphilus* sp.: *Ephedrus plagiator* (Nees)

Padus racemosa — *Rhopalosiphum padi* (L.): *Ephedrus plagiator* (Nees), *Praon abjectum* (Hal.), *Trioxys angelicae* (Hal.)

Philadelphus coronarius — *Aphis fabae* Scop.: *Ephedrus plagiator* (Nees), *Praon abjectum* (Hal.), *Trioxys angelicae* (Hal.)

Quercus spp. — *Tuberculoides annulatus* (Htg.): *Praon flavinode* (Hal.), *Trioxys pallidus* (Hal.)

Ribes grossularia — *Hyperomyzus lactucae* (L.), *Ephedrus plagiator* (Nees)

Ribes nigrum — *Cryptomyzus ribis* (L.): *Aphidius ribis* Hal., *Hyperomyzus lactucae* (L.): *Ephedrus plagiator* (Nees)

Ribes rubrum — *Cryptomyzus ribis* (L.): *Aphidius ribis* Hal.

Robinia pseudoacacia — *Aphis cracivora* (Koch): *Praon abjectum* (Hal.), *Trioxys acalephae* (Marsh.)

Salix spp. — *Aphis farinosa* Gmel.: *Praon abjectum* (Hal.), *Lysiphlebus ambiguus* (Hal.), *Trioxys acalephae* (Marsh.)

Sambucus nigra — *Aphis sambuci* L.: *Praon abjectum* (Hal.), *Trioxys angelicae* (Hal.)

Sorbus aucuparia — *Dysaphis sorbi* (Kalt.): *Ephedrus plagiator* (Nees), *Ephedrus persicae* Frog.

Spiraea spp. — *Aphis spiraephaga* Müller: *Trioxys angelicae* (Hal.), *Trioxys acalephae* (Marsh.), *Ephedrus plagiator* (Nees)

Tilia spp. — *Eucallipterus tiliae* (L.): *Praon flavinode* (Hal.), *Trioxys pallidus* (Hal.)

Ulmus carpinifolia — *Schizoneura ulmi* (L.): *Areopraon lepelleyi* (Wat.), *Ephedrus plagiator* (Nees)

Ulmus effusa — *Tinocallis platani* (Kalt.): *Trioxys hortorum* Starý

Viburnum opulus — *Aphis viburni* Scop.: *Praon abjectum* (Hal.), *Trioxys angelicae* (Hal.), *Ceruraphis eriophori* (Walk.): *Ephedrus plagiator* (Nees)

4. Groups of shrub in fields (field thickets). From the above mentioned viewpoint this type of habitat represents a miscellany of elements of forests, parks and orchard fauna. Under this kind of habitat groups of various shrubs, trees and plants, evolutionarily or artificially developed in borders of woods, pasture meadows, fields, pathways, etc. are included. They serve at first as refugiums for various mammals and birds but their significance for insects, both noxious and useful is rather important, too.

Crataegus monogyna, oxyacantha — *Dysaphis* spp.: *Ephedrus plagiator* (Nees), *Ephedrus persicae* Frog., *Trioxys angelicae* (Hal.)

Euonymus europaea — *Aphis fabae* Scop.: *Ephedrus plagiator* (Nees), *Praon abjectum* (Hal.), *Trioxys angelicae* (Hal.)

Malus silvestris — *Aphis pomi* Deg.: *Trioxys angelicae* (Hal.), *Dysaphis devecta* (Walk.); *Ephedrus plagiator* (Nees), *Ephedrus persicae* Frog.

Pirus communis — *Dysaphis* spp.: *Ephedrus persicae* Frog.

Prunus spinosa — *Brachycaudus* spp.: *Lipolexis gracilis* Först., *Lysiphlebus fabarum* (Marsh.), *Hyalopterus pruni* (Geoffr.): *Ephedrus plagiator* (Nees), *Praon volucre* (Hal.)

Rhamnus cathartica — *Aphis nasturtii* (Kalt.): *Ephedrus plagiator* (Nees)

Robinia pseudoacacia — *Aphis craccivora* (Koch): *Praon abjectum* (Hal.), *Trioxys acalephae* (Marsh.)

Rosa spp. — *Macrosiphum rosae* (Hal.): *Aphidius rosae* Hal., *Praon volucre* (Hal.), *Ephedrus plagiator* (Nees), *Myzaphis rosarum* (Kalt.): *Ephedrus minor* Stelf., *Passerinia tetrarhoda* Walk.: *Ephedrus minor* Stelf.

Viburnum opulus — *Aphis viburni* Scop.: *Praon objectum* (Hal.), *Trioxys angelicae* (Hal.), *Ceruraphis eriophori* (Walk.): *Ephedrus plagiator* (Nees)

5. Orchards

Malus silvestris — *Aphis pomi* (Deg.): *Trioxys angelicae* (Hal.), *Dysaphis* spp.: *Praon volucre* (Hal.): *Dysaphis devecta* (Walk.): *Ephedrus plagiator* (Nees), *Ephedrus persicae* Frog., *Trioxys angelicae* (Hal.)

Pirus communis — *Dysaphis* spp.: *Ephedrus plagiator* (Nees), *Ephedrus persicae* Frog.

Prunus avium — *Myzus cerasi* (F.): *Lipolexis gracilis* Först., *Ephedrus cerasicola* Starý, *Ephedrus plagiator* (Nees), *Ephedrus persicae* Frog.

Prunus cerasus — *Myzus cerasi* (F.): *Ephedrus persicae* Frog.

Prunus domestica — *Brachycaudus cardui* (L.): *Lipolexis gracilis* Först., *Ephedrus persicae* Frog., *Hyalopterus pruni* (Geoffr.): *Ephedrus plagiator* (Nees), *Praon volucre* (Hal.), *Phorodon humuli* (Schrk.): *Ephedrus plagiator* (Nees), *Ephedrus persicae* Frog.

Prunus persica — *Brachycaudus helichrysi* (Kalt.): *Lipolexis gracilis* Först., *Ephedrus persicae* Frog., *Trioxys angelicae* (Hal.)

Ribes grossularia — *Hyperomyzus lactucae* (L.): *Ephedrus plagiator* (Nees)

220

6. Fields

Brassica napus — *Brevicoryne brassicae* (L.): *Diaeretiella rapae* (M'Int.)

Brassica oleracea et var. — *Brevicoryne brassicae* (L.): *Diaeretiella rapae* (M'Int.)

Medicago sativa — *Acyrthosiphon pisum* (Harris): *Aphidius ervi* (Hal.), *Praon dorsale* (Hal.), *Aphis craccivora* (Koch): *Lipolexis gracilis* Först., *Lysiphlebus fabarum* (Marsh.)

Onobrychis viciaefolia — *Aphis craccivora* (Koch): *Lysiphlebus fabarum* (Marsh.)

Sinapis sp. — *Brevicoryne brassicae* (L.): *Diaeretiella rapae* (M'Int.)

Trifolium sp. — *Acyrthosiphon pisum* (Harris): *Aphidius ervi* Hal.

7. Meadows, pastures, waste places, etc.

In this type of habitat belong all the habitats of field type that are not under the direct influence activities of man, ploughing, etc.).

Arctium spp. — *Aphis fabae* Scop.: *Praon abjectum* (Hal.), *Trioxys angelicae* (Hal.), *Brachycaudus cardui* (L.): *Lipolexis gracilis* Först., *Lysiphlebus fabarum* (Marsh.), *Brachycaudus* sp.: *Paralipsis enervis* (Nees), *Chomaphis* sp.: *Lysiphlebus fabarum* (Marsh.), *Paralipsis enervis* (Nees)

Borago officinalis — *Aphis fabae* Scop.: *Ephedrus plagiator* (Nees)

Campanula spp. — *Aphis fabae* Scop.: *Lysiphlebus fabarum* (Marsh.), *Trioxys angelicae* (Hal.), *Dactynotus campanulae* (Kalt.) et spp.: *Praon dorsale* (Hal.), *Trioxys centaureae* (Hal.), *Aphidius funebris* Mack.

Carduus spp. — *Aphis fabae* Scop.: *Lysiphlebus fabarum* (Marsh.), *Brachycaudus cardui* (L.): *Lipolexis gracilis* Först., *Lysiphlebus fabarum* (Marsh.), *Dactynotus aeneus* HRL.: *Aphidius funebris* Mack., *Trioxys centaureae* (Hal.), *Ephedrus campestris* Starý

Centaurea spp. — *Dactynotus jaceae* (L.) et spp.: *Praon dorsale* (Hal.), *Trioxys centaureae* (Hal.), *Macrosiphoniella stägeri* HRL.: *Praon dorsale* (Hal.), *Ephedrus campestris* Starý

Cichorium intybus — *Dactynotus cichorii* (Koch): *Praon dorsale* (Hal.), *Trioxys centaureae* (Hal.), *Aphis intybi* (Koch): *Lipolexis gracilis* Först., *Lysiphlebus fabarum* (Marsh.)

Cirsium arvense — *Aphis fabae* Scop.: *Lipolexis gracilis* Först., *Lysiphlebus fabarum* (Marsh.), *Trioxys angelicae* (Hal.), *Dactynotus* spp.: *Trioxys centaureae* (Hal.)

Crepis biennis — *Dactynotus cichorii* (Koch): *Praon dorsale* (Hal.), *Trioxys centaureae* (Hal.), *Aphidius funebris* Mack., *Ephedrus campestris* Starý

Daucus carota — *Cavariella* spp.: *Trioxys brevicornis* (Hal.), *Aphis lambersi* (Börner): *Lysiphlebus fabarum* (Marsh.), *Dysaphis crataegi* (Kalt.): *Paralipsis enervis* (Nees)

221

Echinops sphaerocephalus — *Paczoskia major* Börner: *Lysiphlebus fabarum* (Marsh.)

Echium vulgare — *Brachycaudus mordwilkoi* HRL.: *Lipolexis gracilis* Först.

Eryngium campestre — *Aphis* sp.: *Lysiphlebus fabarum* (Marsh.)

Melilotus albus — *Acyrthosiphon pisum* (Harris): *Aphidius ervi* Hal., *Therioaphis* sp.: *Praon exoletum* (Nees)

Onobrychis viciaefolia — *Aphis craccivora* (Koch): *Lipolexis gracilis* Först.

Salvia nemorosa — *Aphis salviae* (Walk.): *Lipolexis gracilis* Först.

Salvia pratensis — *Aphis salviae* (Walk.): *Lipolexis gracilis* Först., *Lysiphlebus fabarum* (Marsh.), *Trioxys acalephae* (Marsh.)

Sonchus oleraceus — *Hyperomyzus lactucae* (L.): *Aphidius sonchi* Marsh., *Lysiphlebus fabarum* (Marsh.), *Praon volucre* (Hal.)

Taraxacum officinale — *Aphis taraxacicola* (Börner): *Lipolexis gracilis* Först., *Lysiphlebus fabarum* (Marsh.)

Trifolium spp. — *Acyrthosiphon pisum* (Harris): *Aphidius ervi* Hal.

Vicia spp. — *Acyrthosiphon pisum* (Harris): *Aphidius ervi* Hal., *Praon dorsale* (Hal.), *Aphis craccae* (L.): *Lipolexis gracilis* Först., *Lysiphlebus fabarum* (Marsh.), *Lysiphlebus fritzmuelleri* Mack., *Trioxys acalephae* (Marsh.), *Megoura viciae* Bckt.: *Aphidius megourae* Starý, *Praon dorsale* (Hal.)

Estimation of the aphidiids as parasites of honey plant aphids.

At the estimation special heed was given to the food specificity of individual species, the principal criterion being if a certain parasitic species attacks indifferent or pest aphids. (The food specificity of parasites and other data are given in chapter IV, where pest aphids are marked with a cross.)

The augmentation of the following species should be supported through the aphidofauna of honey plants: *Ephedrus plagiator* (Nees), *Ephedrus persicae* Froggatt, *Lipolexis gracilis* Förster, *Lysiphlebus fabarum* (Marshall), *Praon abjectum* (Haliday), *Praon dorsale* (Haliday), *Praon volucre* (Haliday), *Trioxys angelicae* (Haliday).

The following honey plants with indifferent aphidofauna are recommendable: *Epilobium* spp., *Salix* spp., *Sarothamnus scoparius, Sorbus aucuparia, Viburnum* sp., *Caragana arborescens, Sambucus nigra, Spiraea* sp., *Campanula* sp., *Centaurea* sp., *Cichorium intybus, Plantago* sp.

Some of the plants are weeds, but they do not cause any damage to fallow lands, and in cultures are reliably destroyed by herbicides.

A scheme of the integrated control program.

In our opinion, the scheme of the integrated control of the pest aphids is the following:

1. Aphid species: Life history, foci in environment, population dynamics.

2. Natural limitation of the aphid by parasites (and other enemies): specific composition of parasites, life history, food specificity, effectiveness, foci in the environment.

222

3. Possibilities of the augmentation of native or introduced parasites: Periodic colonization of parasites, artificial inoculation of the host at times of low density, modification of the environment (refuges), introduction of a covercrop, greater plant heterogeneity, possibility of selective breeding.

4. Insecticides. Selection of insecticides, insecticidal treatment at suitable periods when parasites would not be attacked (if the insecticide is not entirely selective, possibility of strip treatment, etc.).

Summary

This book includes results of several years research of aphid parasites in Czechoslovakia. Prior to the publishing of this book a number of smaller previous papers of the author have been published. Nevertheless, the complete elaboration of the mentioned problems was necessary to cover the entire problematics because of too scattered smaller papers that can be obtained only with greater or smaller difficulties as it is usual in such cases. A number of chapters were elaborated in quite an original way.

The purpose of this book is to serve to any entomologist as a certain textbook of the study of aphid parasites. For this reason a certain number of general chapters were included in the book, results of the author's experience of several years of field work. Methods of collecting, preservation and identification of material are dealt with. After a short elaboration of morphology and anatomy the problems of taxonomy are included. In every species the synonymy, distribution, habitat, host, host-specificity, occurrence, are solved to give the general characteristics of every species that can be used in further research, e. g. in using a certain species as aphid control agent. Key to the genera, subgenera and species is added. All the European genera have been keyed to enable at least the generic identification of any aphidiid parasite. As for the key to species, only the species ascertained in Czechoslovakia are included in such keys; for the identification of doubtful species under the generic names mentioned revisions are necessary.

In the chapter on the distribution of species the general principles of the faunistic connections between the aphid parasite fauna of Czechoslovakia and neighbouring countries are dealt with.

The general knowledge of the life history of aphidiid species is rather important for the breeding of every species in the laboratory. Similarly, the knowledge of factors influencing the seasonal occurrence of aphidiids is of great value in the study of both natural limitation and control of aphids by parasites.

Several years samples of material in various kinds of habitats and bred from

224

various hosts in connection with theoretical studies and laboratory observations enables the elaboration of host and parasite interrelationship. This is a rather important chapter, acceptable in a certain way in other groups of entomophagous insects too, being necessary to understand the factors influencing the parasite specificity, host and parasite relationship in ontogeny and phylogeny. The general problem of unnatural host propagation, which is one of the future problems in the research of aphidiid wasps, is also dealt with.

Aphid parasites and aphid attending ants relationship has been studied in quite an original way and various kinds and viewpoints on this relationship have been shown.

Natural enemies of parasites are rather important in influencing the parasite effectiveness. A review of the main groups of the plant and animal kingdom is added.

The research of aphid parasite foci in nature has been originated by the author. The main principles of research, terminology, various types of foci, etc. have been reviewed.

For the purpose of economic entomology pest aphids in agriculture and forestry and their aphid parasites has been reviewed. In every pest aphid species the main features of the life cycle are added in connection with parasitization of various parasite complexes in different types of habitats.

A host and parasite catalogue includes original data, results of several years breeding of parasites from various aphid species in Czechoslovakia.

The chapter on natural limitation and control of aphids by parasites is a result of the book for economic entomologists too. The importance of parasites as natural limitation agents of aphids is discussed. The main principles of using the *Aphidiidae* in biological control of aphids are reviewed in connection with experiments made in Czechoslovakia.

The general chapters of the book may be used at least in all Europe, the keys and ecological characteristics of species are believed to be valuable for the greatest part of Europe too.

References

In this chapter the main taxonomic papers dealing with the European fauna of the *Aphidiidae,* to a lesser degree other papers on world fauna, have been included. The aim of this chapter is to give a list of references, where further more detailed information may be obtained.

Arthur D. E., 1944: Aphidius granarius Marsh. in relation to its control of Myzus kaltenbachi Schout. — Bull. ent. Res. 35 : 257—270.
— 1945: A note on two braconids in their control of corn aphides. — Ent. mon. Mag., London 81: 43—45.
— 1945: The development of artificially introduced infestations of Aphidius granarius Marsh. under field conditions. — Bull. ent. Res. 36: 291—295.
Barnes O. L., 1960: Establishment of imported parasites of the spotted alfalfa aphid in California. — J. econ. Ent. 53: 1094—1096.
Baudyš H. F., 1931: Soupis drobných Hymenopter, které jsem vypěstoval z hálek. — Folia entomol., Brno 4: 60.
Beirne B. P., 1942: Observations on the life-history of Praon volucre Haliday, a parasite of the Mealy Plum Aphis (Hyalopterus arundinis Fab.). — Proc. R. ent. Soc., London (A) 17: 42—47.
— 1942: Observations on the developmental stages of some Aphidiinae. — Ent. mon. Mag., London 78: 283—286.
Bosch R. van den, 1956: Parasites of alfalfa aphid: Natural enemies of spotted alfalfa aphid found in search of Europe and Middle East may become established in California. — California Agric. 10: 7—15.
— 1957: Status of imported parasites of the spotted alfalfa aphid (Therioaphis maculata) in California. — Ann. Ent. Soc. Amer. 50: 27.
— 1957: The spotted alfalfa aphid and its parasites in the Mediterranean Region, Middle East, and East Africa. — J. econ. Ent. 50: 352—356.
Bosch R. van den, Schlinger E. I., 1962: Biological control of the spotted alfalfa aphid in southern California. — XI. Int. Ent. Kongr. Wien 1960: 2—47.
Bosch R. van den, Schlinger E. I., Dietrick E. J., Hall J. C., 1959: The role of imported parasites in the biological control of the spotted alfalfa aphid in southern California in 1957. — J. econ. Ent. 52: 142—154.
Bosch R. van den, Schlinger E. I., Dietrick E. J., Hagen K. S., Holloway J. K., 1959: The colonization and establishment of imported parasites of the spotted alfalfa aphid in California. — J. econ. Ent. 52: 136—141.

226

Bosch R. van den, Schlinger E. I., Hagen K. S., 1962: Initial field observations in California on Trioxys pallidus (Haliday) a recently introduced parasite of the walnut aphid. — J. econ. Ent. 55: 857—862.

Dalla Torre C. G., 1898: Catalogus Hymenopterorum huiusque descriptorum systematicus et synonymicus. — Vol. IV. Braconidae. Lipsiae.

Dragoun J., 1933: Biologia lumčíka Aphidius fabarum Marsh., parasitujícího na mšici Aphis fabae Scop. — Cukrovarnické listy, Praha 51: 137—140.

Dunn J. A., 1949: The parasites and predators of potato aphids. — Bull. ent. Res., London 40: 97—122.

Eidmann H., 1924: Die Eiablage von Trioxys Hal. nebst Bemerkungen über die wirtschaftliche Bedeutung dieses Blattlausparasiten. — Z. ang. Ent. 10: 353—363.

Fahringer J., 1937: Die Parasiten der Baumläuse (Lachnini) aus der Gruppe der Aphidiinae Först. — Festschr. 60. Geb. E. Strand, Riga 3: 240—245.

Fedotova K. M., Rjachovsky V. V., 1954: Aphidius ervi Hal. (Aphidiidae) and its importance in pea aphid outbreaks (in Russian). — Trudy Inst. ent. fitop., Kiev 5: 87—90.

Förster A., 1862: Synopsis der Familien und Gattungen der Braconen. — Nerh. naturh. Ver. Rheinl. 19: 225—250.

Gahan A. B., 1911: Aphidiinae of North America. — Bull. Maryland Agr. Expt. Sta. 152: 147—200.

George K. S., 1957: Preliminary investigations on the biology and ecology of the parasites and predators of Brevicoryne brassicae (L.). — Bull. ent. Res. 48: 619—629.

Goidanich A., 1934: Materiali per lo studio degli Imenotteri Braconidi. — II. Boll. Lab. ent. Bologna 6: 209—230.

Griffiths D. C., 1960a: The behaviour and specificity of Monoctonus paludum Marshall (Hym., Braconidae), a parasite of Nasonovia ribis-nigri (Mosley) on lettuce. — Bull. ent. Res. 51: 303—319.

—   1960b: Immunity of aphids to insect parasites. — Nature 187: 346.

—   1961: The development of Monoctonus paludum Marshall (Hym., Braconidae) in Nasonovia ribis-nigri on lettuce, and immunity reactions in other lettuce aphids. — Bull. ent. Res. 52: 147—163.

Hafez M., 1961: Seasonal fluctuations of population density of the cabbage aphid, Brevicoryne brassicae (L.) in the Netherlands, and the role of its parasite, Aphidius (Diaeretiella) rapae (Curtis). Wageningen (Thesis).

Haliday A. D., 1833: An essay on the classification on the parasitic Hymenoptera of Britain, which correspond with the Ichneumones minuti of Linnaeus. — Ent. Mag., London 1: 259—276. Ditto, 1833; 1: 480—491. Ditto, 1834, 2: 93—106.

Hall J. C., Schlinger E. I., Bosch R. van den, 1962: Evidence for the separation of the "sibling species" Trioxys utilis and Trioxys pallidus. — Ann. Ent. Soc. Amer. 55: 566—568.

Haviland M. D., 1918—1919: On the life-history and bionomics of Myzus ribis Linn. (red — — currant aphis). — Proc. R. Soc., Edinburgh 39: 78—112.

—   1920: Preliminary note on a cynipid hyperparasite of aphides. — Proc. Camb. Phil. Soc., Cambridge 20: 235—238.

—   1921: On the bionomics and development of Lygocerus testaceimanus Kieffer and Lygocerus cameroni Kieffer (Proctotrypoidea, Ceraphronidae), parasites of Aphidius. — Quart. Jl. Microsc. Sci. 65: 101—127.

—   1921: On the bionomics and post-embryonic development of certain cynipid hyperparasites of aphides. — Quart. Jl. Microsc. Sci. 65: 451—478.

—   1922: On the post-embryonic development of certain chalcids, hyperparasites of aphides, with remarks on the bionomics of hymenopterous parasites in general. — Quart. Jl. Microsc. Sci. 66: 321—338.

Hazelhoff E. H., 1929: Biological control of the sugar-cane aphid by transferrring its native parasite from the old to young fields. — Trans. 4th Ent. Congr. Ent. 2: 55—61.

Hincks W. D., 1944: On the "shirt-button" cocoon of Dyscritulus planiceps (Marshall). Naturalist, London 810: 93—96.

Hodek I., Holman J., Starý P., Štys P., 1959: Natural enemies of the bean aphid (Aphis fabae Scop.) in Czechoslovakia. — Trans. I. Conf. Insect. Path. and Biol. control, Praha 1958: 553—557.

Hodek I., Starý P., Štys P., 1962: The natural enemy complex of Aphis fabae and its effectiveness in control. — Verh. XI. Int. Kongr. Ent., Wien 1960 2: 747—749.

Ivanova-Kasas O. M., 1961: Note on comparative embryology of Hymenoptera. Leningrad, 265 pp. (in Russian).

Janiszewska J., 1933: Untersuchungen über die Hymenoptere Aphidius sp. Parasiten der Blattlaus Hyalopterus pruni Fabr. — Bull. Int. Acad. Cracovie 1932 2: 277—292.

Johnson B., 1958: Influence of parasitization on form determination in aphids. — Nature, London 181: 205—206.

— 1958: Effect of parasitization by Aphidius platensis Brèthes on the developmental physiology of its host Aphis craccivora Koch. — Ent. expt. et appl. 2: 82—99.

Kirchner L., 1856: Die Ichneumonen der Umgegend von Kaplitz. Prag.

Krombein K. V., et al., 1958: Hymenoptera of America North of Mexico. — Synoptic catalog. U.S. Dept. Agric. Monogr. 2, First Suppl., 305 pp.

MacGill E. I., 1923: The life history of Aphidius avenae (Hal.), a braconid parasite of the nettle aphis (Macrosiphum urticae). — Proc. R. Soc. Edinburgh 43: 51—71.

Mc Leod J. H., 1937: Some factors in the control of the common greenhouse aphid, Myzus persicae Sulzer, by the parasite Aphidius phorodontis Ashm. — Rep. ent. Soc. Ont. 67: 63—64.

Mackauer M., 1958: Zur Kenntnis der paläarktischen Aphidiinae (Hym., Braconidae). 1. Beitrag: Die wirtschaftliche Bedeutung von Aphidius ribis Hal. — Z. ang Ent. 43: 282—285.

— 1959a: Histologische Untersuchungen an parasitierten Blattläusen. — Z. Parasitenkde 19: 322—352.

— 1959b: Die mittel-, west- und nordeuropäischen Arten der Gattung Trioxys Haliday. — Beitr. Ent. 9: 144—179.

— 1959c: Ein entomophager Parasit der Mooslaus. — Dtsch. ent. Z., N. F. 6: 82—85.

— 1959d: Die systematische Stellung von Aphidius pseudoplatani Marsh. Senck. biol., Frankfurt M. 40: 179—182.

— 1959e: Die europäischen Arten der Gattungen Praon und Areopraon. — Beitr. Ent. 9: 810—865.

— 1959f: Trioxys similis n. sp. (Hym., Brac., Aphidiinae) eine neue Blattlaus — Schlupfwespe aus Frankreich. Nebst einigen biocönologischen und nomenklatorischen Bemerkungen. — Entomophaga 4: 303—309.

— 1960a: Zur Systematik der Gattung Trioxys Haliday. — Beitr. Ent. 10: 137—160.

— 1960b: Die europäischen Arten der Gattung Lysiphlebus Förster. — Beitr. Ent. 10: 582—623.

— 1960c: Ein spezifischer Parasit der Blattlausgattung Holcaphis HRL. — Boll. Lab. Ent. Agr. Portici 18: 294—297.

— 1960d: Ein weiterer Vertreter der Untergattung Betuloxys Mackauer. — Mitt. Dtsch. Ent. Ges. 19: 96—98.

— 1961a: Die Typen der Unterfamilie Aphidiinae des Britischen Museums London. — Beitr. Ent. 11: 96—154.

— 1961b: Neue europäische Blattlaus-schlupfwespen. — Boll. Lab. Ent. Agr. Portici 19: 270—290.

— 1961c: Zur Frage der Wirtsbindung der Blattlausschlupfwespen. — Z. Parasitenkde 20: 576—591.

— 1961d: Spezifische Parasiten der schwarzen Blattläuse und verwandter Arten. — Mitt. Biol. Bundesanst. Berlin-Dahlem, 104.

— 1961e: Die Gattungen der Familie Aphidiidae und ihre verwandschaftliche Zuordnung. — Beitr. Ent. 11: 792—803.

— 1962a: Blattlaus- Schlupfwespen der Sammlung F. P. Müller, Rostock. — Beitr. Ent. 12: 631—661.

— 1962b: Wirtsbindung der Aphidiinae und Fahrenholzsche Regel. XI. — Int. Ent. Kongr. Wien 1960 2: 733—738.

— 1962c: Spezifische Parasiten der Acyrthosiphon — Macrosiphum Gruppe und Grundfragen der Wirtsbindung der Blattlaus — Schlupfwespen. — Z. ang. Ent. 50: 125—131.

Maneval H., 1940: Observations sur un Aphidiidae myrmécophile. Description du genre et de l'espece. — Bull. Soc. Linn. Lyon 9: 9—14.

Marshall T. A., 1896, 1897: Braconides, in André: Species des Hyménoptères d'Europe et d'Algerie. Vol. 5,5 bis.

Muesebeck C. F. W., 1956: Two new parasites of the yellow clover aphid and the spotted alfalfa aphid. — Bull. Brooklyn ent. Soc. 51: 25—28.

Muesebeck C. F. W., and Walkley L. M., 1951: in Muesebeck C. F. W., Krombein K. V., Townes H. K., 1951: Hymenoptera of America North of Mexico. — Synoptic catalog. U.S. Dept. Agric. Monogr. 2.

Narayanan E. S., Subba Rao B. R., Sharma A. K., 1960: A catalogue of the known species of the world belonging to the subfamily Aphidiinae. — Beitr. Ent. 10: 545—581.

Narayanan E. S., Subba Rao B. R., Sharma A. K., Starý P., 1962: Revision of "A catalogue of the known species of the world belonging to the subfamily Aphidiinae". — Beitr. Ent. 12: 662—720.

Nees v. Esenbeck, 1818: Genera et familias Ichneumonidum adscitorum exhibens. — Verh. Leop. Carol. Acad. Naturf. Erlangen 9: 299—310.

— 1834: Hymenopterorum Ichneumonibus affinium monographiae, genera europaea et species illustrantes. Stuttgartiae et Tubingiae.

Niezabitowski E. L., 1909: Materyaly do fauny Brakonidów Polski. — Spraw. Kom. Fizyogr. Kraków, 44: 47—106.

Obrtel R., 1961: Effects of two insecticides on Aphidius ervi Hall (Hym., Brac.), an internal parasite of Acyrthosiphon onobrychis (Boyer). — Folia zoologica, Brno 10: 1—8.

Pimentel D., 1961: Natural control of aphid populations on cole crops. — J. econ. Ent. 54: 885—888.

Pontin A. J., 1960: Some records of predators and parasites adapted to attack aphids attended by ants. — Ent. mon. Mag., London 95: 154—155.

Quilis M. P., 1929: Estudio biologico del ichneumonido Aphidius avenae Hal., parásito de los pulgones verdes. — EOS, Madrid 5: 427—459.

— 1931: Especies nuevas de Aphidiidae españoles. — EOS, Madrid 7: 25—84.

— 1934: Algunos Aphidiidae de Checoslovaquia. — EOS, Madrid 10: 5—19.

Ratzeburg J. T. C., 1844, 1848, 1852: Die Ichneumonen der Forstinsekten. Berlin.

Schimitschek E., 1935: Forstschädlingsauftreten in Österreich 1927 bis 1933. — Zentralbl. ges. Forstwes. 61: 134—221.

Schlinger E. I., 1960a: Diapause and secondary parasites nullify the effectiveness of rose — aphid parasites in Riverside, California, 1957—1958. — J. econ. Ent. 53: 151—154.

— 1960b: The latest import direct from France. — Diamond Walnut News 42: 9—10.

— 1960c: Natural enemies of aphids. Handb. of biological control of plant pests. — Plants and gardens 16: 36—42.

Schlinger E. I., Dietrick E. J., 1960: Biological control of insect pests aided by strip-farming alfalfa in experimental program. — Calif. agric. 14: 8.

Schlinger E. I. Hagen K. S., Bosch R. van den, 1960: Imported French parasite of walnut aphid established in California. — Calif. agric., Berkeley 14: 3—4.

Schlinger E. I. Hall J. C., 1959: A synopsis of the biologies of three imported parasites of the spotted alfalfa aphid. — J. econ. Ent., 52: 154—157.

— 1960a: Biological notes on pacific coast aphid parasites, and lists of California parasites (Aphidiinae) and their aphid hosts. — Ann. Ent. Soc. Amer. 53: 404—415.

— 1960b: The biology, behavior and morphology of Praon palitans Muesebeck, an internal parasite of the spotted alfalfa aphid, Therioaphis maculata (Buckton). — Ann. Ent. Soc. Amer. 53: 144—160.

— — 1961: The biology, behavior and morphology of Trioxys utilis, an internal parasite of the spotted alfalfa aphid, Therioaphis maculata. — Ann. Ent. Soc. Amer. 54: 34—45.

Sedlag U., 1958: Beobachtungen über das Uftreten der Kohlblattlaus (Brevicoryne brassicae L.) im Sommer 1957. — Nachrbl. Dtsch. Pflzschtzdnst. 12: 73—77.

Sekhar P. S., 1957: Mating, oviposition and discrimination of hosts by Aphidius testaceipes (Cresson) and Praon aguti Smith, primary parasites of aphids. — Ann. Ent. Soc. Amer. 50: 370—375.

— 1959: Life-history of Aphidius testaceipes (Cresson) and Praon aguti (Smith) — Hym. Braconidae — primary parasites of aphids with notes on the effects of parasitism on hosts. — Current Sci. 28: 333—335.

— 1960: Host relationships of Aphidius testaceipes (Cresson) and Praon aguti (Smith), primary parasites of aphids. — Canad. J. Zool. 38: 593—603.

Short J. R. T. 1952: The morphology of the head of larval Hymenoptera with special reference to the head of Ichneumonoidea, including a classification of the final instar larvae of the Braconidae. — Trans. R. ent. Soc., London 103: 27—84.

Skriptshinsky G., 1930: Zur Biologie von Aphidius granarius Marsh. und Ephedrus plagiator (Nees) (Braconidae), Parasiten von Aphis padi L. (in Russian with German summary). — Rep. App. Ent., Leningrad 4: 351—364.

Skuhravý V., Novák K., Starý P., 1959: Entomofauna des Kleefeldes (Trifolium pratense L.) und ihre Entwicklung. — Rozpravy ČSAV, Praha, MPV 69 (7): 84 pp.

Smith C. F., 1944: The Aphidiinae of North America. — Ohio State Univ. Contr. Zoo. Ent., Columbus 6: 154 pp.

Spencer H, 1926: Biology of the parasites and hyperparasites of aphids. — Ann. Ent. Soc. Soc. Amer. 19: 119—157.

Starý P., 1958: A taxonomic revision of some aphidiine genera with remarks on the subfamily Aphidiinae. — Acta Faun. Ent. Mus. Nat. Pragae 3: 53—96.

— 1959a: Redescription of the aphidiine genus Lipolexis Förster, 1862. — Acta Soc. ent. Čechoslov. 56: 93—96.

— 1959b: A revision of the European species of the genus Monoctonus Haliday. — Acta Soc. ent. Čechoslov. 56: 237—250.

— 1959c: A revision of the genus Dyscritulus Hincks. — Acta Faun. Ent. Mus. Nat. Pragae 5: 69—74.

— 1959d: Notes on Aphidius ephippium Hal. — Beitr. Ent. 9: 180—184.

— 1959e: Some problems connected with the research of Aphidiinae (Hym., Brac.), as natural enemies of aphids with regard to their utilization for the purpose of biological control. — Trans. I. Conf. Insect Path. and Biol. control, Praha 1958: 537—541.

— 1960a: Une nouvelle espece du genus Trioxys Hal. de la région de la Pannonie (Tchécoslovaquie). — Bull. Soc. ent. Mulhouse 1960: 93—96.

230

— 1960b: The generic classification of the family Aphidiidae. — Acta Soc. ent. Čechoslov. 57: 238—252.

— 1960c: A taxonomic revision of the European species of the genus Paraphidius Starý, 1958. — Acta Faun. Ent. Mus. Nat. Pragae 6: 5—44.

— 1960d: Two new species of Trioxys Haliday from Central Europe. — Acta Soc. ent. Čechoslov. 57: 365—368.

— 1960e: The aphidiid genus Lysaphidus Smith C. F. in Europe. — Bull. ent. Pologne 30: 357—366.

— 1961a: Notes on European species of the genus Aphidius Nees. — Ent. Tidskr. 82: 213—221.

— 1961b: Notes on the parasites of the root aphids. — Acta Soc. ent. Čechoslov. 58: 228—238.

— 1961c: A revision of the genus Diaeretiella Starý. — Acta Ent. Mus. Nat. Pragae 34: 383—397.

— 1961d: Faunistic survey of Czechoslovak species of the genera Lysiphlebus Förster and Trioxys Haliday. — Acta Faun. Ent. Mus. Nat. Pragae 7: 131—149.

— 1961e: Taxonomic notes on the genus Lysiphlebus Förster. — Bull. ent. Pologne 31: 97—103.

— 1961f: Two new species of Praon Haliday from Czechoslovakia. — Acta Soc. ent. Čechoslov. 58: 340—343.

— 1962a: Aphidofauna of honey plants as a source of subsidiary hosts of aphidiid wasps. — — Acta Soc. ent. Čechoslov. 59: 42—58.

— 1962b: Aphid parasites (in Czech). — Živa, Praha 10: 62—63.

— 1962c: New aphid parasites of the genus Aphidius Nees from Europe. — Bull. ent. Pologne 32: 109—122.

— 1962d: Hymenopterous parasites of the pea aphid Acyrthosiphon onobrychis (Boyer) in Czechoslovakia. — Folia zoologica, Brno 11: 265—278.

— 1962e: Notes on European species of the genus Ephedrus Haliday. — Opusc. entomol. 27: 87—98.

— 1962f: Bionomics and ecology of Ephedrus pulchellus Stelfox, an important parasite of leaf-curling aphids in Czechoslovakia, with notes on the diapause. — Entomophaga 7: 91—100.

— 1963a: A study on the relationship of the Dactynotinae and their aphidiid parasites in Europe. — Acta Ent. Mus. Nat. Pragae 35: 593—610.

— 1963b: A study on the relationship of the Myzinae and their aphidiid parasites in (Central) Europe. — Boll. Lab. Ent. Agr. Portici 21: 199—216.

— 1964a: Food specificity in the Aphidiidae. — Entomophaga 9: 91—99.

— 1964b: Integrated control problems of citrus and peach aphid pests in Italy orchards. — Entomophaga 9: 147—152

— 1964c: Biological control of Megoura viciae Bckt. in Czechoslovakia. — Acta Soc. ent. Čechosl. 61: 301—322

— 1964d: The foci of aphid parasites (Hymenoptera, Aphidiidae) in nature. — Ekol. Polska
— A 12: 529—554

— in press: Aphid parasites of aphids in the USSR. — Acta Faun-ent. Mus. Nat. Pragae

— A study on the relationship of the Pterocommatinae and the Aphidinae, and their aphidiid parasites in (Central) Europe. — Acta Ent. Mus. Nat. Pragae

— inp ress: A study on the relationship of the Anuraphidina and their aphidiid parasites in Europe. — Beitr. Ent.

Starý P., Rupais A., 1963: The parasites of dendrophilous aphids in East Baltic (in Russian). — Latv. Entom., Riga 7: 63—67.

Starý P., Sedlag U., 1959: Aphidius (Metaphidius) trioxyformis, eine neue Art und Untergattung der Aphidiinae. — Dtsch. ent. Z. 6: 160—165.

Stern V. M., Smith R. F., Bosch R. van den, Hagen K. S., 1959: The integration of chemical and biological control of the spotted alfalfa aphis. — Hilgardia 29: 81—101.

Szépligeti V., 1898: Beiträge zur Kenntnis der Ungarischen Braconiden. — III. Termes. Füzet. 21: 381—408.

Telenga A. N., 1950: On the utilization of aphidiid parasites in migrant aphids control. — Nautsh. Tr. inst. ent. phytop. AN USSR, Kiev 2: 199—209.

Thomson C. G., 1895: LII. Bidrag till Braconidernas kännedom. — Opusc. entomol. 20: 2141—2339.

Vevai E. J., 1942: On the bionomics of Aphidius matricariae Hal., a braconid parasite of Myzus persicae Sulz. — Parasitology, London 34: 141—151.

Way M. J., 1963: Mutualism between ants and honey-dew producing Homoptera. — Ann. Rev. Ent. 8: 307—344.

Wheeler E. W., 1923: Some braconids parasitic on aphids and their life-history. — Ann. Ent. Soc. Amer. 16: 1—29.

Wiackowski S. K., 1961: Laboratornye studia nad biologia i ekologia pasorzytnicej blonkowki Aphidius smithi Sharma and Subba Rao (Hym., Braconidae) sprowadzonej z Pakistanu do Kalifornii do biologiczengo zwalcania mszycy grochowej — Acyrthosiphon pisum (Harr.). — Postepy Nauk Roln., Z. Probl. 35: 137—141.

—   1962: Studies on the biology and ecology of Aphidius smithi Sharma and Subba Rao (Hymenoptera, Braconidae), a parasite of the pea aphid, Acyrthosiphon pisum (Harr.). — Bull. ent. Pologne 32: 253—310.

Williams F. X., 1931: Handbook of the insects and other invertebrates of Hawaiian sugar cane fields. Adv. Publ. House — Honolulu, 400 pp.

Zwölfer H., 1958: Zur Systematik, Biologie und Ökologie unterirdisch lebender Aphiden. — Z ang. Ent. 43: 1—52.

Note: The literature on aphids has been omitted in this list of references.

Explanation of Plates

Plate I

All figures are drawn from female specimens except where otherwise stated.

1. Nomenclature of thorax, dorsal view. TG — tegulae, AANT — fore wings, APST-hind wing, PN — pronotum, NOT-notaulices, MSCT — mesoscutum, AX — axillae, SCTL — scutellum, MTNT — metanotum, PSCTL — postscutellum, PROP — propodeum, SP — spiracles, $T_{1,2}$ — first and second tergite. 2. Nomenclature of thorax, lateral view: Abbreviations see Fig. 1. PN — pronotum, PP — propleurae, $CX_{1,2,3}$ — coxae, MSPL — mesopleurae, MTPL — metapleurae. 3. Nomenclature of wing venation. C — costal vein, Sc — subcostal vein, Pt — pterostigma, Mt — metacarpus, Ptc — pterostigmal cell, R — radial vein, $Rc_{1,2,3}$ — radial cells 1, 2, 3, $Ir_{1,2}$ — interradial veins, 1, 2, Bc — basal cell, B — basal vein, Mc — median cell, M — median vein, Im — intermedian vein, Cu — cubital vein, $Cuc_{1,2}$ — cubital cell, 1, 2, An — anal vein, n — nervulus. 4. Nomenclature of last instar larva of *Aphidius megourae*. 5. Female genitalia (after Smith C. F., 1944). MP — median prong, PTGR — proctiger, VP — ventral prong, IX — tergite 9, 2VLF — second valvifer, 3VL — valvulae 3 (= ovipositor sheaths), 1VLF — valvifer 1, AP — anterior prong, 1VL — valvulae 1,2VL — valvulae 2. 6. Nomenclature of male genitalia. GX — gonocoxit, GB — gonobasis, PV — penis valvae, P — penis, GR — gonostipital rami, D — digitus, CS — cuspis, GF — gonoforceps, APD — apodems of penis valvae. 7. Nomenclature of head and distances. OC — ocellus, interocul. — interocular line, FR — frons, socket ocul. — socket ocular line, facial — facial line, head w. — head width, clypeoant. — clypeoantennal line, transfac. — transfacial line, FC — face, tent-ocul. — tentorio — ocular line, GE —gena, MD — madible, CL — clypeus, intertent. — intertentorial line, 0 — eye.

Plate II

8. *Ephedrus plagiator* (Nees), instar I larva (after Ivanova Kasas). 9. *Aphidius megourae* Starý, egg. 10. *Aphidius megourae* Starý, instar I larva. 11. *Aphidius megourae* Starý, instar II larva. 12. *Aphidius megourae* Starý, instar III larva. 10—12 drawn in relative dimensions. 13. *Lipolexis gracilis* Förster, instar I larva. 14. *Monoctonus angustivalvus* Starý, instar I larva. 15. *Lysiphlebus melandriicola* Starý, wing. 16. *Lysiphlebus salicaphis* (Fitch), wing. 17. *Lysiphlebus ambiguus* (Haliday), wing. 18. *Diaeretiella rapae* (M'Intosh), wing.

Plate III

19. *Diaeretellus ephippium* (Hal.), male wing. 20. *Trioxys angelicae* (Hal.), wing. 21. *Ephedrus plagiator* (Nees), wing. 22. *Ephedrus persicae* Frog., wing. 23. *Lipolexis gracilis* Förster, wing. 24. *Aphidius rosae* Hal., wing. 25. *Dyscritulus planiceps* (Marsh), wing.

Plate IV

26. *Trioxys pannonicus* Starý, wing. 27. *Ephedrus lacertosus* Hal., wing. 28. *Monoctonus angustivalvus* Starý (and 30), variation in wing venation. 29. *Diaeretus leucopterus* (Hal.), wing. 30. *Monoctonus angustivalvus* Starý, wing. 31. *Praon* sp., wing. 32. *Paralipsis enervis* (Nees), wing.

Plate V

33. *Pauesia abietis* (Marsh.), genitalia. 34. *Aphidius rosae* Hal., genitalia. 35. *Areopraon lepelleyi* (Waterston), genitalia. 36. *Pauesia laricis* (Hal.), genitalia. 37. *Aphidius caraganae* Starý, propodeum. 38. *Trioxys falcatus* Mackauer, genitalia. 39. *Trioxys parauctus* Starý, tergite. 1. 40. *Diaeretus leucopterus* (Hal.), genitalia. 41. *Trioxys macroceratus* Mackauer, genitalia. 42. *Trioxys parauctus* Starý, genitalia.

Plate VI

43. *Lysaphidus arvensis* Starý, genitalia. 44. *Trioxys centaureae* Hal., genitalia. 45. *Dyscritulus planiceps* (Marsh.), genitalia. 46. *Ephedrus cerasicola* Starý, propodeum. 47. *Lysaphidus arvensis* Starý, head from above. 48. *Aclitus obscuripennis* Först., head from above. 49. *Lysaphidus erysimi* Starý, head from above. 50. *Pauesia picta* (Hal.), genitalia. 51. *Trioxys pannonicus* Starý, genitalia. 52. *Aphidius avenae* Hal., propodeum. 53. *Lysiphlebus fabarum* Marsh., genitalia. 54. *Monoctonus angustivalvus* Starý, propodeum. 55. *Ephedrus plagiator* (Nees), genitalia. 56. *Monoctonus caricis* Hal., propodeum. 57. *Pauesia unilachni* (Gahan), genitalia.

Plate VII

58. *Trioxys heraclei* (Hal.), genitalia. 59. *Paralipsis enervis* (Nees), head, lateral view. 60. *Trioxys hortorum* Starý, genitalia. 61. *Dyscritulus planiceps* (Marsh.), head from above. 62. *Toxares deltiger* (Hal.), genitalia. 63. *Diaeretellus ephippium* (Hal.), genitalia. 64. *Trioxys auctus* (Hal.), genitalia. 65. *Pauesia picta* (Hal.), mesoscutum, lateral view. 66. *Monoctonus crepidis* (Hal.), propodeum. 67. *Diaeretus leucopterus* (Hal.), propodeum. 68. *Ephedrus validus* (Hal.), propodeum. 69. *Diaeretiella rapae* (M'Int.), genitalia. 70. *Praon volucre* (Hal.), genitalia. 71. *Monoctonus crepidis* (Hal.), genitalia. 72. *Aphidius rosae* Hal., propodeum. 73. *Lysiphlebus thelaxis* Starý, tergite 1. 74. *Lysiphlebus fabarum* (Marsh.), tergite 1. 75. *Ephedrus brevis* Stelfox, tergite 1.

Plate VIII

76. *Pauesia maculolachni* (Starý), head, frontal view. 77. *Trioxys humuli* Mackauer, genitalia. 78. *Aclitus obscuripennis* Först., abdomen. 79. *Monoctonus angustivalvus* Starý, genitalia. 80. *Me-*

taphidius aterrimus (Fahr.), abdomen. 81. *Metaphidius aterrimus* (Fahr.), propodeum. 82. *Monoctonia pistaciaecola* Starý, genitalia. 83. *Lysaphidus arvensis* Starý, tergite 1. 84. *Areopraon lepelleyi* (Waterston), propodeum. 85. *Lysiphlebus thelaxis* Starý, tibia. 86. *Lysiphlebus arvicola* Starý, tergite 1. 87. *Praon pubescens* Starý, flagellar segments. 88. *Praon volucre* (Hal.), flagellar segments. 89. *Pauesia picta* (Hal.), propodeum. 90. *Monoctonia pistaciaecola* Starý, tergite 1. 91. *Trioxys angelicae* (Hal.), genitalia. 92. *Ephedrus nacheri* (Quilis), tergite 1.

Plate IX

93. *Pauesia picta* (Hal.), tergite 1. 94. *Trioxys angelicae* (Hal.), tergite 1. 95. *Pauesia laricis* (Hal.), tergite 1. 96. *Trioxys auctus* (Hal.), tergite 1. 97. *Pauesia pini* (Hal.), tergite 1. 98. *Ephedrus plagiator* (Nees), apex of flagellum, 99. *Trioxys centaureae* (Hal.), tergite 1. 100. *Lysiphlebus dissolutus* (Nees), propodeum. 101. *Trioxys heraclei* (Hal.), tergite 1. 102. *Ephedrus minor* Stelfox, apex of flagellum. 103. *Ephedrus validus* (Hal.), genitalia. 104. *Trioxys macroceratus* Mackauer, tergite 1. 105. *Pauesia cupressobii* (Starý), tergite 1. 106. *Ephedrus validus* (Hal.), flagellar segments 1. 2. 107. *Pauesia unilachni* (Gahan), mesoscutum, lateral view. 108. *Trioxys pallidus* (Hal.), genitalia. 109. *Lipolexis gracilis* Förster, genitalia. 110. *Trioxys spinosus* Starý, genitalia. 111. *Trioxys glaber* Starý, genitalia. 112. *Trioxys betulae* (Marsh.), genitalia.

Plate X

1. *Megoura viciae* Bckt. colonies heavily infested by *Aphidius megourae* Starý. Laboratory rearings. 2. *Aphidius megourae* Starý, male. 3. *Megoura viciae* Bckt. colony heavily infested by *Aphidius megourae* Starý, detailed. 4. *Hyalopterus pruni* (Geoffr.) colony on peach leaf, heavily infested by *Aphidius transcaspicus* Tel. (Sicily).

Plate XI

5. *Byrsocrypta ulmi* (L.) on *Ulmus campestris*. 6. *Phyllaphis fagi* (L.) on *Fagus silvatica*. 7. *Myzus cerasi* (F.) on *Prunus avium*. 8. *Cryptomyzus ribis* (L.) on *Ribes rubrum*.

Plate XII

9. *Aphidius megourae* Starý, final instar larva removed from mummified aphid's skin. 10. *Aphidius megourae* Starý, final instar larva inside the aphid skin, beginning of mummification. 11. Mummified *Megoura viciae* Bckt. aphid by its parasite *Aphidius megourae* Starý. 12. Mummified *Megoura viciae* Bckt. aphids with emergence holes of *Aphidius megourae* Starý.

Plate XIII

13. *Trioxys angelicae* Hal. infesting lower instar aphids (*Aphis fabae* Scop.) in a colony. 14. *Aphis fabae* Scop., parasitized and mummified by *Lysiphlebus fabarum* (Marsh.). 15. Cocoon of *Praon abjectum* (Hal.), 16. Cocoon of *Dyscritulus planiceps* (Marsh.) with alate adult aphid (*Drepanosiphum platanoides* Schrk.) on its top.

Plate XIV

17. Ant runs around root collar of *Scabiosa columbaria*, with a colony of *Aphis thomasi* (Börn) inside. 18. Mummified (black) aphids (*Brachycaudus mordwilkoi* HRL.) on root collar of *Carduus* sp., parasitized by *Paralipsis enervis* (Nees). 19. Pemphigine aphids on roots of plants in an ant nest. 20. *Sorbus aucuparia*, leaves curled by *Dysaphis sorbi* (Kalt.).

Plate XV

21. Galls on *Ulmus* sp., caused by *Schizoneura lanuginosa* Htg. 22. Galls on *Artemisia absinthium*, caused by *Cryptosiphum artemisiae* Bckt. 23. *Hyalopterus pruni* (Geoffr.) colony of *Prunus* sp. 24. *Aphis schneideri* Börn. colony protected by *Formica* sp. ant on *Ribes rubrum*.

Plate XVI

25. *Eriosoma lanigerum* (Hausm.) colonies on *Malus silvestris* 26. Diapause-cocoon of *Ephedrus persicae* Froggatt. 27. Ditto, inside dried curled leaf of *Malus silvestris*. 28. Mummified *Sitobium* sp. aphids on *Avena sativa*, parasitized by *Aphidius avenae* Hal. 29. *Nasonovia nigra* HRL. on *Hieracium silvaticum*, parasitized and mummified by *Monoctonus angustivalvus* Starý.

Plate XVII

30. *Microsiphum millefolii* Wahlgr. on *Achillea millefolium*, protected by *Formica* sp. ants. 31. Submountain meadows with *Juniperus communis*, Vsetínské vrchy. 32. Beech forest with undergrowth of *Hieracium* sp., etc., Javorník. 33. Vineyards of eastern Slovakia, in close neighbourhood of which *Prunus amygdalus*, *Prunus avium*, *Prunus persica*, etc. are grown.

Plate XVIII

34. Wood steppe *(Quercus pubescens)*, eastern Slovakia. 35. Groups of shrubs in fields *(Sambucus nigra, Euonymus europaea, Prunus spinosa)*, southern Moravia. 36. Sugar beet field and a balk, where foci of bean aphid parasites can be found. Environs of Prague. 37. Southern Moravia landscape, an example of a cultivated district, where numerous foci of parasites may be found due to the existence of balks where more or less a natural community still exists.

Plate XIX

38. Peat-bog, southern Bohemia. 39. Steppe preservation, env. of Mikulov, southern Moravia. 40. Roadsides (chronic foci of parasites) in cultivated district, southern Slovakia. 41. Heavily infested alfalfa field by *Acyrthosiphon pisum* (Harris), where the pest aphid was sucessfully limited by *Aphidius ervi* Hal.

Plate XX

42. A railway station environs, fallow land grown by weeds *(Artemisia vulgaris, A. absinthium)*, an example of chronic indifferent parasite focus. 43. A heap of manure in fields, grown by *Cirsium*

arvense and other weeds. A chronic focus of aphidiid parasites, of *Lysiphlebus fabarum* (Marsh.) namely. 44. Potato field with *Chenopodium album* weeds. The weed represents temporary foci of useful parasite *Diaeretiella rapae* (M'Int.), a parasite of *Hayhurstia atriplicis* L. on *Chenopodium*. This parasite infests also *Myzodes persicae* Sulz. on potato. 45. Laboratory glass used for rearing of *Megoura viciae* Bckt. and other aphids on a single plant. 46. Laboratory mass rearing of aphids and parasites under fluorescent light.

Plate XXI

47. Mass-production of aphids and parasites on bean plants in a greenhouse. 48. Aspirator-suction collector with exchangeable plastic bottles, used for mass-collection of parasites in a greenhouse. 49. Liberation of adult parasites and parasitized aphids in the field.

PLATE I

PLATE II

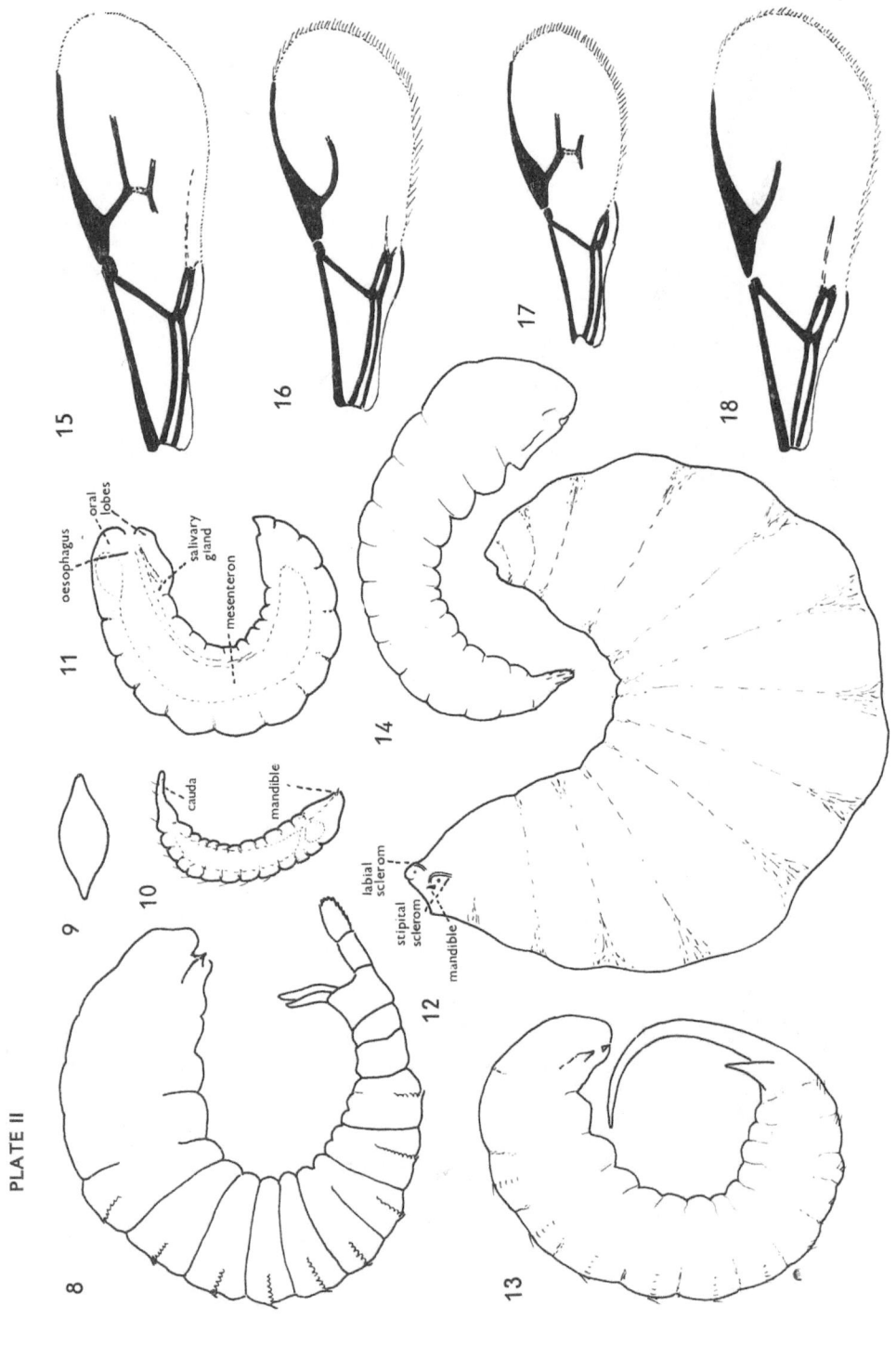

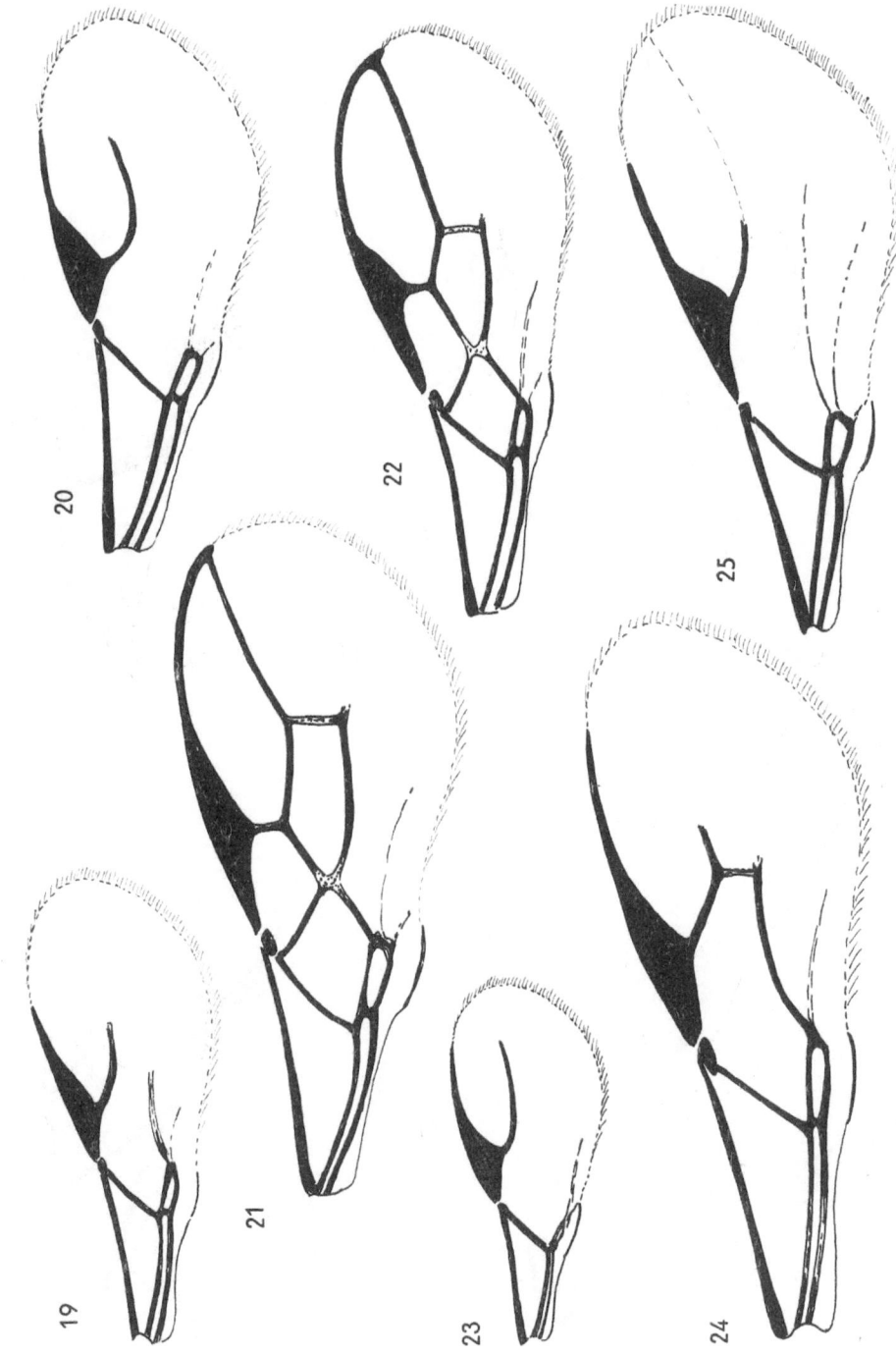

PLATE III

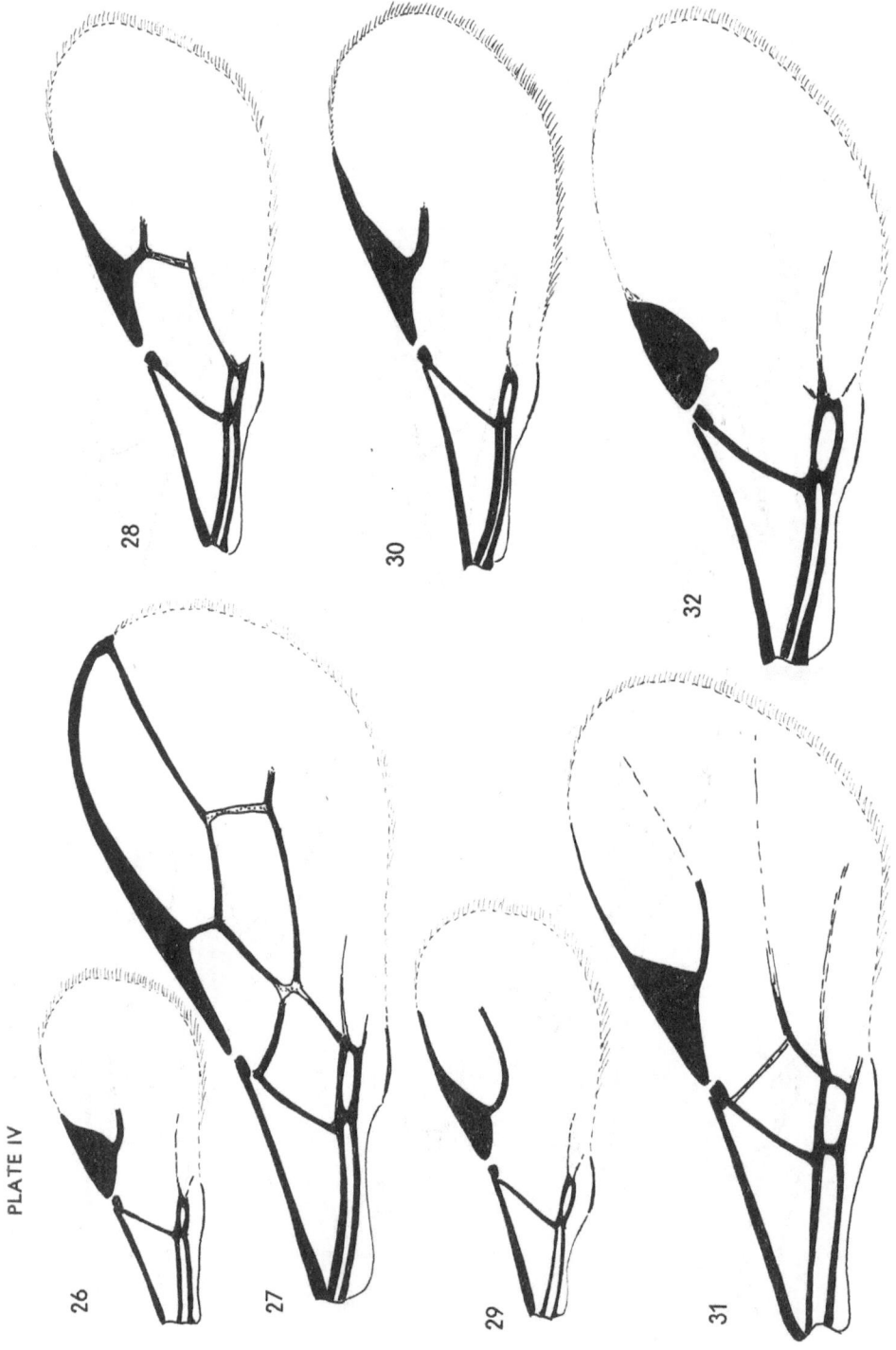

PLATE IV

PLATE V

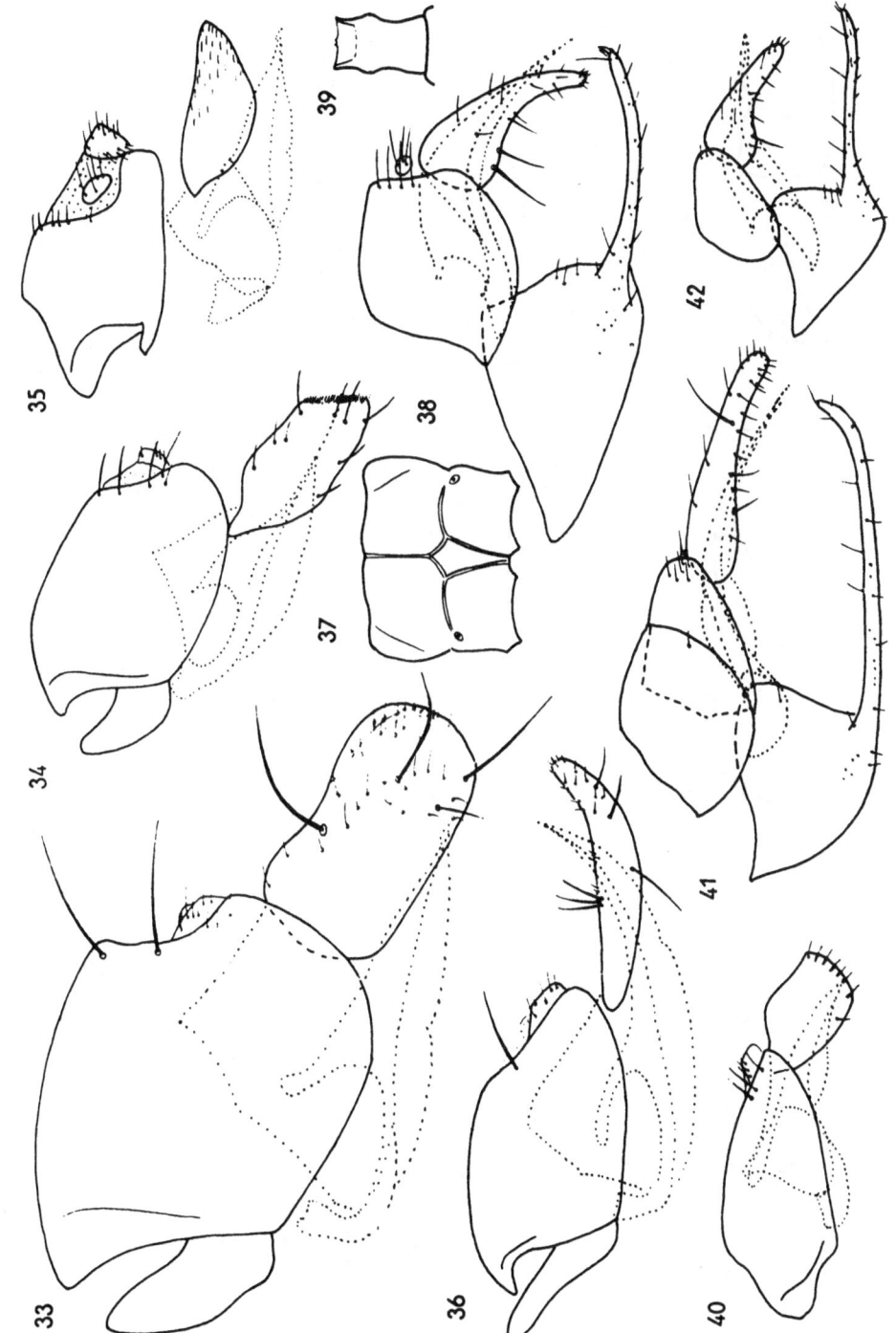

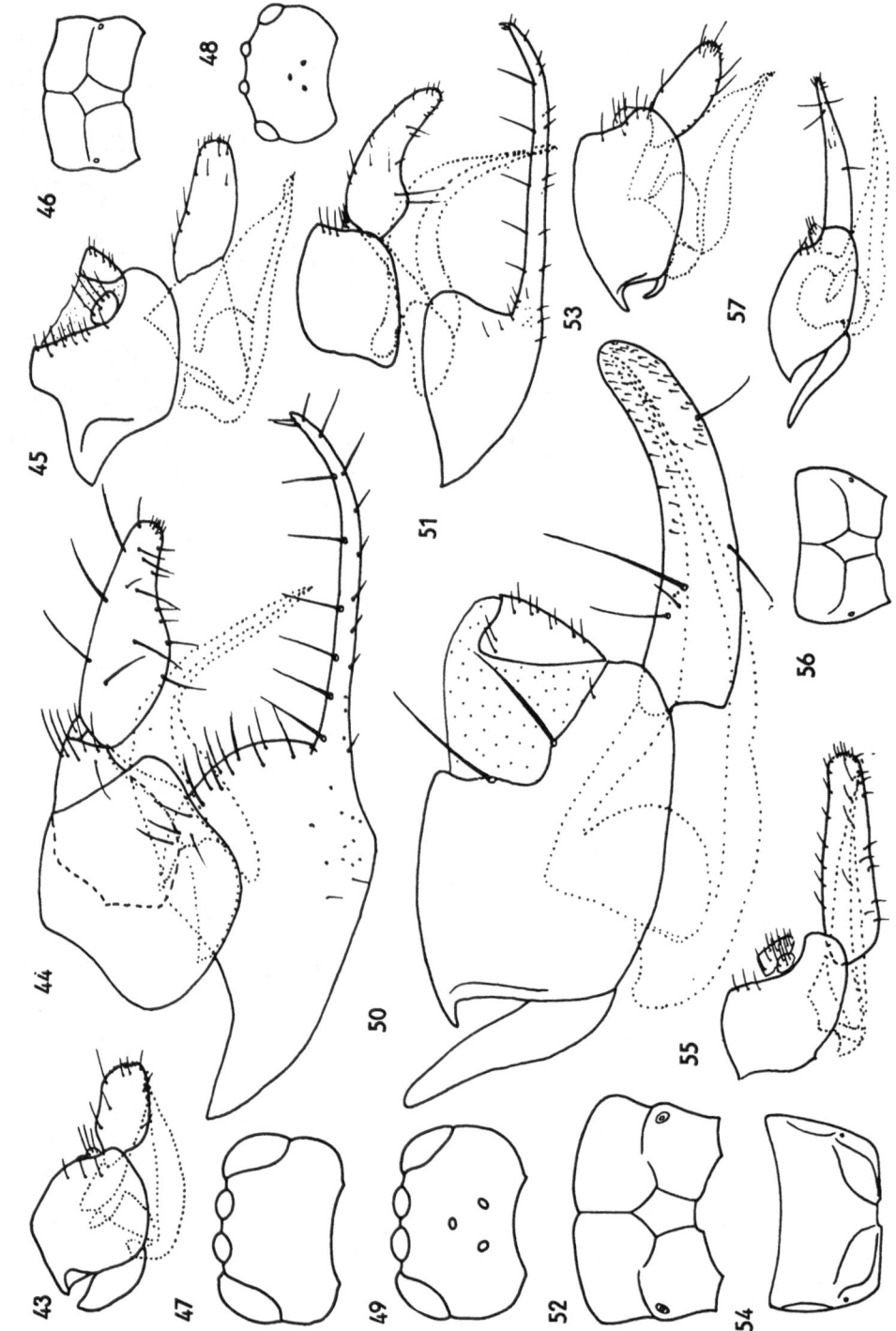

PLATE VI

PLATE VII

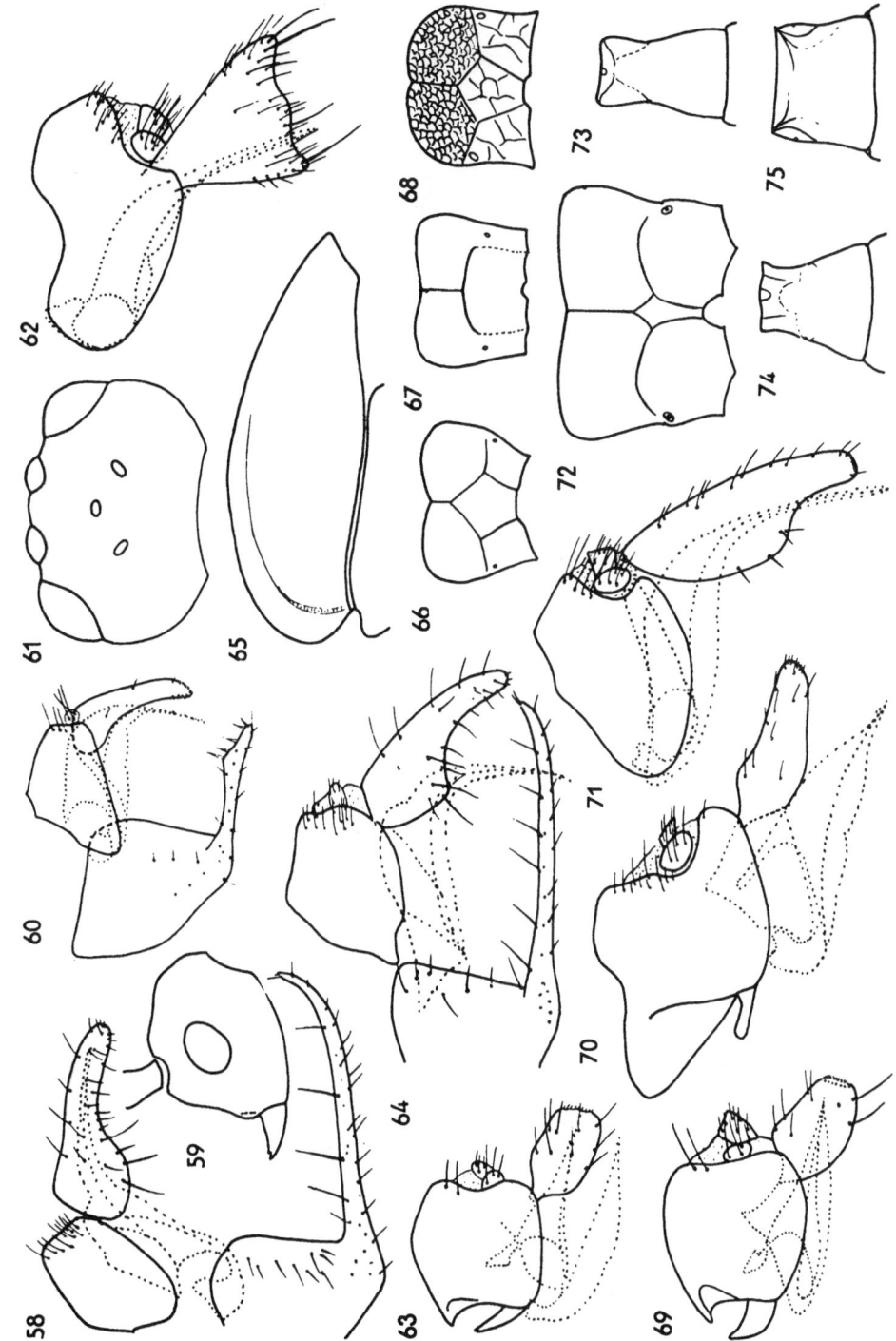

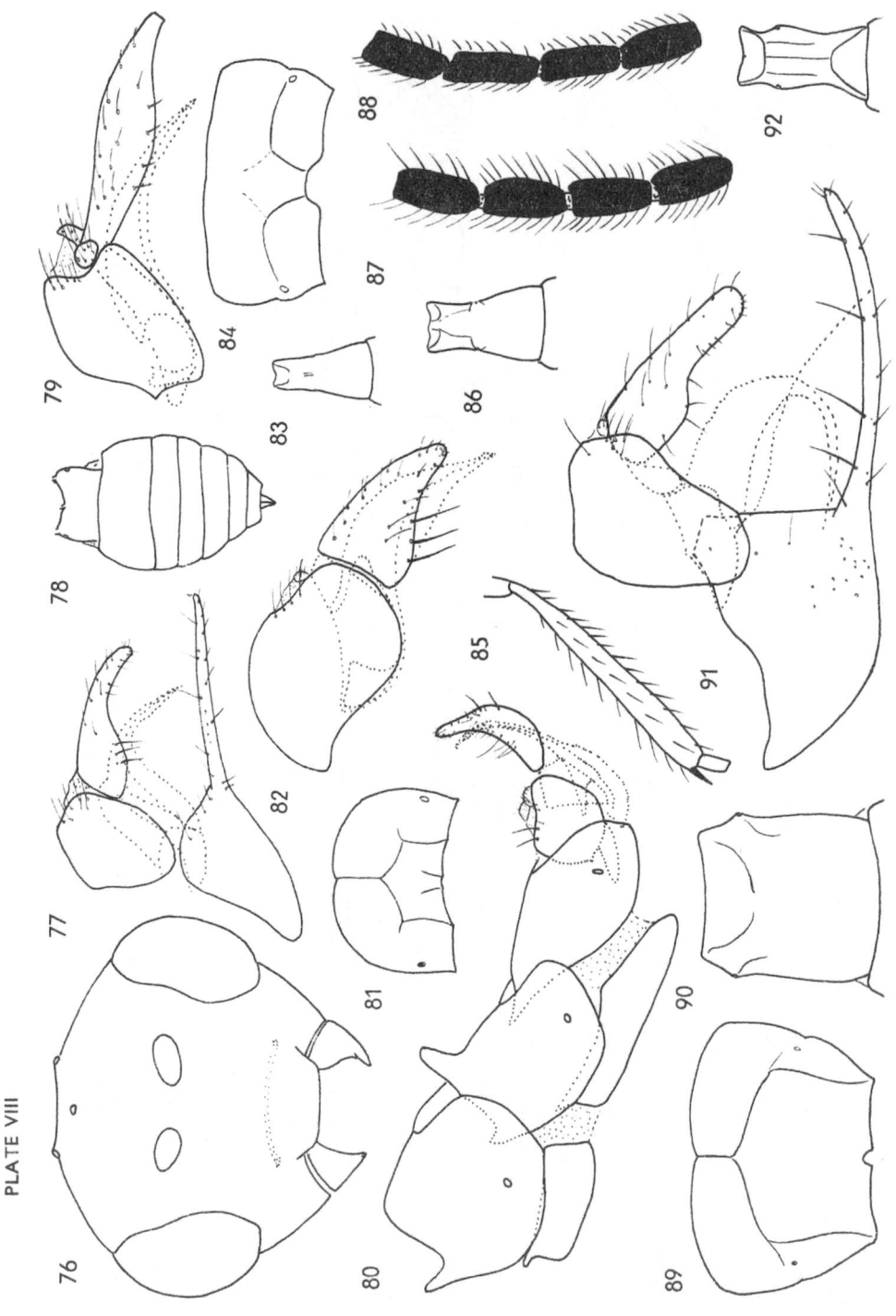

PLATE VIII

PLATE IX

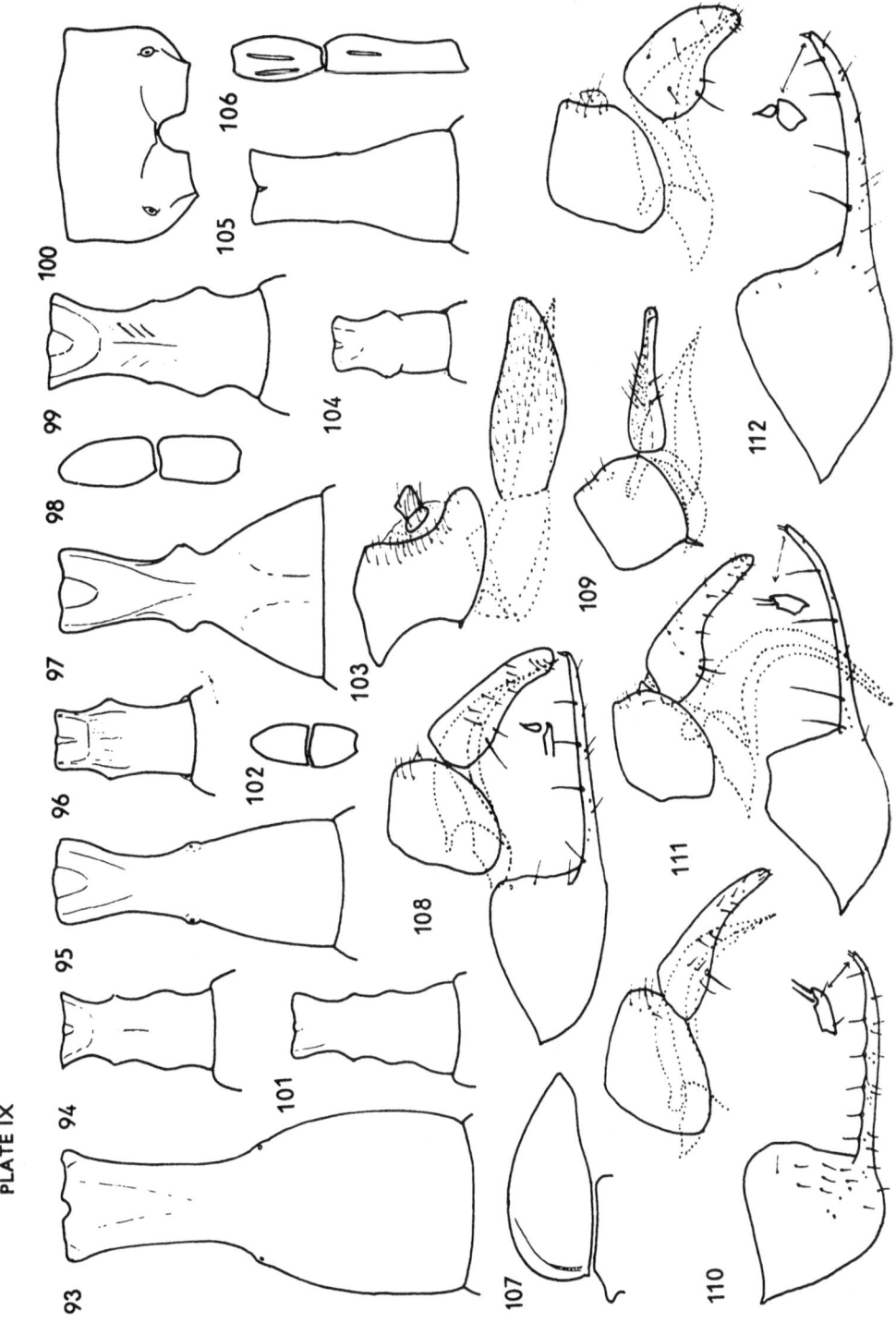

PLATE X

PLATE XI

PLATE XII

PLATE XIII

PLATE XIV

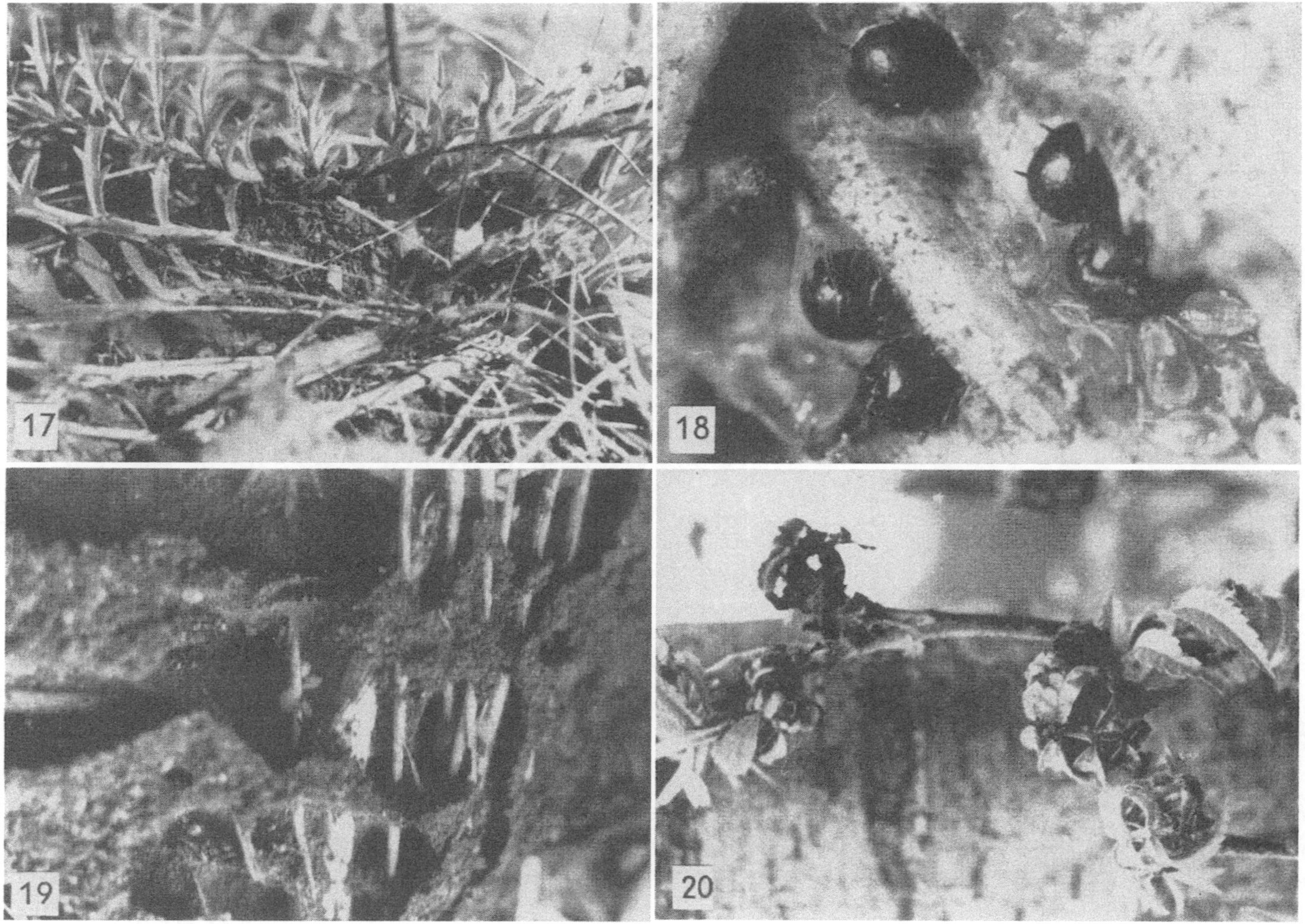

17

18

19

20

PLATE XV

PLATE XVI

25

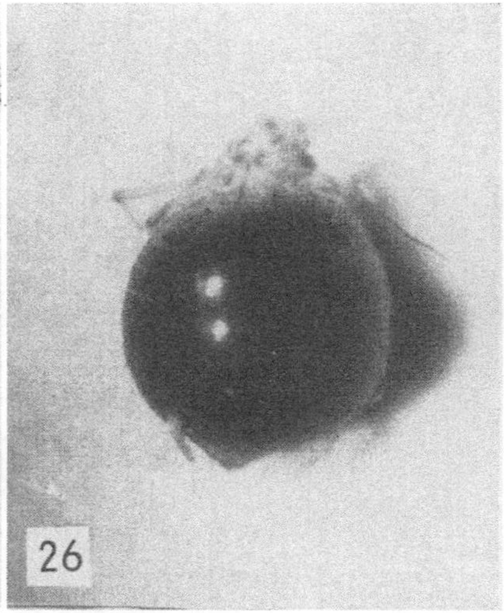

26

27

28

29

PLATE XVII

30

31

32

33

PLATE XVIII

34

35

36

37

PLATE XIX

PLATE XX

PLATE XXI

47

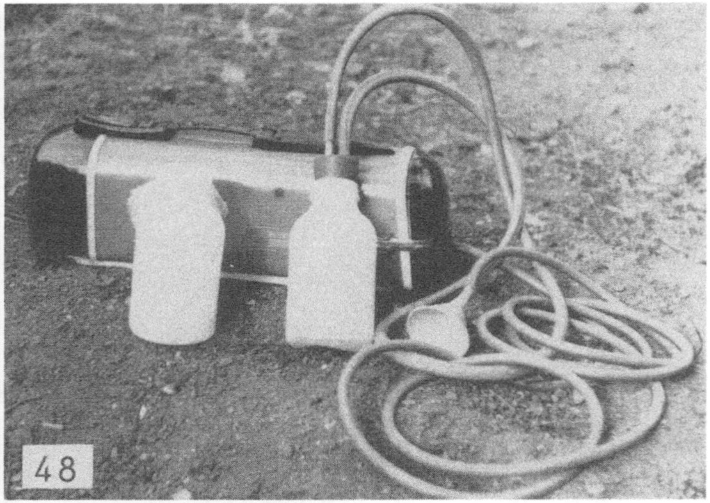

48

49

Index

Notes: Host aphids records are not included in the Index. For these records, see: Host and parasite catalogue.

Synonyms of parasite species are omitted.

240

241

242